THE RISE AND FALL

—— OF THE ——

AMERICAN TEENAGER

ALSO BY THOMAS HINE

The Total Package
Facing Tomorrow
Populuxe

THE RISE AND FALL

—— OF THE ——

AMERICAN TEENAGER

Thomas Hine

Perennial

An Imprint of HarperCollins*Publishers*

A hardcover edition of this book was published in 1999 by Bard,
an imprint of Avon Books, Inc.

HarperCollins books may be purchased for educational, business, or sales promotional use.
For information please write: Special Markets Department, HarperCollins Publishers Inc.,
10 East 53rd Street, New York, NY 10022.

First Perennial edition published 2000.

Designed by Kellan Peck

The Library of Congress has catalogued the hardcover edition as follows:

Hine, Thomas
The rise and fall of the American teenager / Thomas Hine.—1st ed.
p.   cm.
ISBN 0-380-97358-8
1. Teenagers—United States. 2. Adolescence—United States. I. Title.
HQ796.H493   1999   99-24381
305.235'0973—dc21   CIP

ISBN 0-380-72853-2 (pbk.)

08 RRD 10 9 8

For my aunt, Genevieve Dolan,
*a lifelong teacher who always
listened seriously to young people.*

CONTENTS

THE RISE AND FALL

—— OF THE ——

AMERICAN TEENAGER

INTRODUCTION
Are Teenagers Necessary?

Several months after beginning this book, I had a sudden realization that I had actually begun it three decades before.

I was sixteen, just beginning my senior year in high school. Debbie, a close friend of mine who was also the literary editor of the yearbook, came to me with an assignment. The yearbook committee, good baby boomers that they were, had decided to dedicate that year's edition neither to a teacher that had guided generations of students through the hazards of polynomials, nor to the late President Kennedy, but to "the spirit of youth." In other words, we were dedicating it to us. After all, had anyone else really been young before, in quite so intense a way?

She had found people to write about laughter and study and friendship and sport. "I have a topic I think you'd be just perfect for," she said, as have so many editors since, hoping that my self-evident suitability for an unappetizing topic would overcome any resentment or insult I might feel. "I want you to do the trials of youth," she said, her voice rising with enthusiasm. "The problems of youth, the misery." Or at least that's how I remember it. I went home and wondered for a day or two whether I was so obviously the most miserable of my peers. Then I wrote the piece.

All these years later, I dug into a closet in my mother's house and unearthed the 1965 edition of *Menunketuck*. The piece was addressed to people who, like me, would look at the yearbook in middle age. It cautioned against being nostalgic for these youthful years. Its chief metaphor involved birds breaking their necks by flying into plate-glass windows, which was certainly gloomy enough to keep Debbie off my back.

It was written as an interior monologue of a young person wondering about the Big Issues:

"Maybe I'm something special, and maybe I'm not. Maybe I'm here for a reason and I might be going somewhere after this, but then again I might not. I wonder where I fit in?"

By the time I read those sentences again, I had read a lot of studies that indicated that I'd put my finger on something. Figuring out where they fit in—to the universe, the world, the economy, their social circle, their family—is a project on which teenagers spend a lot of their time and energy. Despite the mythology of youth as a revolutionary and utopian time, study after study suggests that teenagers' principal preoccupation is to adapt, to find a place in life.

But the most valuable thing I found in returning to my yearbook wasn't this long-ago statement of the obvious. I found myself. I didn't Find Myself, as adolescents are expected to do. Rather, I found a voice that had been there all along—my voice, my feelings, my way of thinking. I found myself as a beginner, practicing a craft that I'd learn better, making some mistakes that I wouldn't make now.

Still, this kid—I—really did start this book at that moment. And there's been no change of authors.

All of us have been teenagers, and we ought to be experts on how teenagers think. Oddly, few people can actually remember the experience. Anna Freud theorized that the experience of adolescence is so filled with pain, trauma, and turmoil that our conscious minds suppress it. There's a serious problem with this theory, though. Survey after survey of young people show that they aren't miserable at all. They have problems, of course, but they feel confident about coping with them.

My theory on why we don't remember ourselves as teenagers is simpler: It's simply that we remember ourselves as ourselves. Being a teenager isn't an identity but a predicament most people live through. And typical teenagers are people we don't know very well. I don't know anyone who thinks she was one herself.

This book isn't about my inner teenager, or yours either. Rather, it's about people in their teens throughout American history—what adults expected of them, how they fit into the world they found, and how they helped shape it as well. There's no room here for nostalgia—an emotion that distorts understanding both of the past and of the present. The sixties, the decade of my teenage years, have been celebrated—and demonized—quite enough.

So this is the last you'll hear about that sixteen-year-old would-be writer. Yet he'll be sitting with me as I write, making a nuisance of himself. His role is to keep reminding me that while this is largely a history of roles and expectations, the teenagers I'm discussing aren't some exotic species—they're real people. And that's what makes them so difficult to deal with.

This book grew out of a certain exasperation with my generation. We seem to have moved, without skipping a beat, from blaming our parents for the ills of society to blaming our children. We want them to embody virtues we only rarely practice. We want them to eschew habits we've never managed to break. Their transgressions aren't their own. They send us the unwelcome, rarely voiced message that we, the adults, have failed.

I'm not a parent, but as I've been working on the book, I've had plenty of offers of help from people who are. "You can have mine to study," they say, adding that they'll take the kid back in three or four years. They love their children and look forward to their recovery from teenagerhood. They're baffled and often enraged by that sullen creature doing God knows what in the bedroom upstairs.

There's no doubt that teenage offspring often make parents' lives difficult. And maybe only a nonparent would set out to write a book that has, as one of its themes, the way that society makes young people's lives difficult. But if our conception of teenaged lives is dysfunctional, parents suffer along with their children. So even though this isn't a "how to deal with your troubling teen" book, perhaps both parents and their sons and daughters can learn something about the predicament they're in.

I began working on this book with a suspicion—which matured into a conviction—that what's failing is neither the older generation nor the younger. The problem is, rather, a way of thinking about an important time of life and how people ought to spend it. The problem is with the very concept of "teenager."

The word was coined during the early 1940s by some anonymous writer or editor to describe an age group that had suddenly become of great interest to marketers and social reformers. Like the Hoover Dam, the American teenager was a New Deal project, a massive redirection of energy. The national policy was to get the young out of the workforce so that more jobs would be available to family men. For the first time, high schools were enrolling a majority of young people of high school age. During the late 1930s, Mickey Rooney provided a convincing and engaging model of arrested development. And as a final, necessary ingredient, the buildup to a wartime economy provided teenagers with pocket money. Money plays a paradoxical role for teenagers. If they are in the mainstream workforce, they're not teenagers. But if they don't have any money, no youth culture emerges.

The concept of the teenager rests in turn on the idea of the adolescent as a not quite competent person, beset by stress and hormones. This idea was popularized by the psychologist and educator G. S. Hall about a century ago, but it rests on data and assertions that have not withstood scientific scrutiny.

At most, we can say that the teenager is a social invention, one that took shape during the first half of the twentieth century in response to a society very different from our own. While, for example, a bottle-making machine or a microprocessor is quickly supplanted when a better one is invented, a social invention like the teenager is difficult to displace. It has become embedded in the way in which we imagine the shape of people's lives. It has become part of the cycle that begins at birth and ends at death. Nothing seems more real or more changeless than that.

The purpose of this book is to sketch out a useful context for thinking about the young—one that jettisons obsolete and destructive aspects of our contemporary view of teenagers. The teenager was an invention of the Machine Age. We live in a different world now, and we must reconsider how we think about the lives of the young.

Although I will draw on a number of disciplines—from psychology to anthropology to marketing—the primary emphasis is historical. Without a historical perspective, change—especially change that's so close to home and which seems to challenge fundamental values—can only feel like a threat. Everything seems to be crumbling. Ideas and institutions that appeared true and eternal seem to be under siege, and what is taking their place seems empty, or even evil.

★   ★   ★

The point of looking back is not to observe that everything that happens is for the best. History is far too horrifying to support that point of view. What looking back does tell us is that all sorts of things can and do happen. It shows us that the shape of people's lives has varied tremendously over time. It shows us that even the family, that institution we so often identify as the bedrock of human life, has changed significantly several times during the relatively short span of American history.

For most of our history, the labor of young people in their teens was too important to be sacrificed. Europeans observed that Americans grew quickly in every way, taking on responsibilities and vices much sooner than their European counterparts. Although some prosperous urban families concluded, during the nineteenth century, that there were greater long-term rewards from investing in their children's training and education than in cashing in on their labor, this remained a minority view well into the twentieth century.

Yet, the concept of the teenager as we know it today is not simply a continuation of the self-consciously sheltered life of the middle-class nineteenth-century high school student. The street workers of the big cities, who made their living selling newspapers, carrying messages, and shining shoes, helped generate a lively and youthful—if crude—popular culture. The youthful horse culture of frontier youths swiftly metamorphosed into the car culture. Immigrants' children taught their parents how to behave in America and thus acquired a measure of authority we still accord our youth.

My story begins way before there was anyone called a teenager or an adolescent, when the word "youth" could describe a person from twelve to thirty-five. Indeed, most of the book is about people in their teens who weren't "teenagers." Only by looking at people in their teens in the centuries before there were teenagers is it possible to understand how artificial the concept of a teenager is.

When I started out, I expected that the book would be mostly about the twentieth century and the development of youth culture. But I quickly became engrossed in the great variety of youthful experience before the teenager began to take shape. The story of people in their teens is not a simple progression, or even a single story. In a 1750s classroom, a nineteen-year-old might have learned from the same textbook as an eight-year-old. And a sixteen-year-old could be a physician! A hundred years later, the teenage girl might be a factory worker, her brother a businessman. A century after that, we would find them all in high school, and a kid in Memphis with a bizarre fashion sense could set styles for the entire world. This range of adult expectations and youthful competence is itself a key lesson for considering what kinds of preparation will best suit teenagers in the future.

"Diversity" has been one of the buzzwords of the last quarter century, though much of the story I will tell here shows how the experiences of young people in America have become far less diverse over time. During the nineteenth century especially, the lives of young people working in the streets in big cities were different from those who were working in mills in smaller cities only a few dozen miles away. The youth of the coal mines of Pennsylvania were breadwinners for their families, while young people involved in mining in the Rockies were independent, and even entrepreneurial. Young people studying at high schools and living with their families experienced a different sort of education than those who boarded near academies. Most youths during this entire period were on farms, though there was a big difference between being part of a family trying to establish itself in the wilderness and one that was farming an established area. And obviously, youth as a slave was an absolutely distinctive experience.

Many of these diverse experiences were horrific, and we're well rid of them. Yet each of these different modes of youth shaped the world we live in today. And young people's success in adapting to so many roles suggests that they may have greater abilities than we give them credit for.

Following an opening chapter about the state of the contemporary teenagers are two chapters that deal with the physiological, psychological, and cultural dimensions of youth. The next four chapters are a roughly chronological account of young people and their place in the family, the economy, and society from the seventeenth century to the turn of the twentieth. The next two chapters trace the origins and development of the high school and of the idea of adolescence. The next five chapters consider the twentieth century, and particularly the phenomena of youth culture and universal education. A concluding chapter looks at the possibility of a world after teenagers.

The last thirty years have brought a wealth of research on the history of American families and young people. My purpose here was not to add to this impressive body of work, but to consider it in light of contemporary issues. My methods are impure. I am quite frankly looking at the past for contemporary relevance, and even the historical chapters incorporate modern parallels. My goal is to use these historical findings in an attempt to imagine the varieties of youthful experience our country has already seen or may yet create. Armed with a richer sense of teen-agers' possibilities, perhaps we can free ourselves from turn-of-the-twentieth-century ideas and conceive of more appropriate, more satis-fying ways of being young.

When I speak of the rise of the teenager, I'm really talking about the acceptance of the idea that youth is a time for experimentation and protracted preparation, usually in school. If there is something heroic about speaking of a rise, that's quite intentional. Both teenagers and their elders tend to believe, at least some of the time, in the romantic idea that to be human, you must become the hero of your own life. Youth is seen, especially by adults, as a time without compromise, when you don't have to become someone you don't respect in order to make a living, please a boss, and meet your obligations to others. This potential nobility becomes mock-heroic, however, when you realize that the young person's quest for self is going to be played out not in Valhalla but in high school. High school, which virtually defines the rise of the teenager, is hardly an exalted place.

For many individuals, such a long period of education, exploration, and deferred responsibility has been a tremendous gift. For other individ-uals, it has not been a blessing. The absence of a significant economic role for young people has made them dependent on their families for longer periods than their ancestors often were. Young people are often judged to be less able than they are. The concept of the teenager has been an impediment that has kept them from becoming the people they were ready to be.

This lengthy waiting period has tended to reduce young people's contacts with older people and increase them with people who are ex-actly the same age. That in turn has led to the rise of a youth subculture that has helped define and elaborate what it means to be a teenager. Any account of the rise of the teenager is, in large part, an account of the changing shape and continuing importance of teen culture.

The rise of the teenager is a straightforward idea, the fall is less so. That's because the fall can be imagined in several different ways.

In one sense, the idea of rise and fall has been implicit in the definition of the teenager from the beginning. The teenager is the symbol of Americans' rising aspirations, the repository of hopes, the one who will realize the American dream. And inevitably, the teenager is a disappointment, whose combination of adult capacities and juvenile irresponsibility sows personal heartbreak and social chaos.

Teenagers have been "falling" one way or another—by dropping out of school, becoming pregnant, joining gangs—from the time society first started raising the age of adulthood.

If you take a romantic view, young people "fall" in another way: They grow up. They cut their hair (or remove their tattoos), give up their youthful idealism, and fade into the gray mass of adult society.

Still another notion of the "fall" is implicit in the fears and complaints that adults have about young people—especially now that teen-bashing is such a popular bipartisan activity. Being a teenager is less and less what Erik Erikson proposed—a moratorium period in which to find your identity. Teenagers are losing their license for irresponsibility while, at the same time, they continue to be denied a role in their society, other than that of style setters and consumers.

Yet, I'd like to think that this "fall" could be a happy one—if our destructive ideas about teenagers give way to ones that grant young people a more decisive role in their own lives.

When Adam and Eve partook of the Tree of Knowledge, they freed themselves from their benevolent but paternalistic creator, and entered into a life that was more difficult, but recognizably human. Similarly, the fall of the idea of teenager might expose young people to new dangers, different ways of failing, and difficult new decisions. But some say that Adam and Eve's fall was a fortunate one—that the greater responsibility they had to bear was, in itself, a form of freedom. So could it be for the post-teenage young. Besides, being a teenager today is no paradise.

A note about terminology:

Standard references cite a 1941 article in *Popular Science* magazine as the first published use of the word "teenager." The term came into use during World War II and first turned up in a book title in 1945. It seems to have leaked into the language from the world of advertising and marketing, where demographic information was becoming an increasingly important part of predicting which sales approaches are most effective with particular buyers.

References to a person in his or her teens had been part of the language for since the 1600s. But such references had always described individuals. With the rise of the persuasion industries during the twentieth century, large groups of people were increasingly identified by single characteristics. People in their teens became "teens" or "teeners" or "teen-agers." They were largely in the same place—high school—sharing a common experience, and they were young and open to new things. They were, in short, easy to sell to. Moreover, the preferences formed when young often endure, which makes selling to teenagers a reasonable long-term investment.

The word "teen" has some interesting obsolete meanings that seem to echo faintly today. For seven centuries, "teen" meant a source of anger, irritation, or anxiety, an often apt description of one's offspring. It also meant barrier, and "teenage" (with a short *a*) was wood long enough for making a high fence—a meaning with resonance for young people who feel that being categorized as a teenager limits their freedom.

While it's anachronistic to talk about teenagers before 1940, I have done so. It is easy, in reading history, to see the people involved as completely unlike ourselves. Yet we make a fetish of classifying all sorts of people purely by chronological age and organizing their lives accordingly. Thus, I will sometimes call these earlier young people teenagers, not to deny the uniqueness of their experience, but to remind you that if they were living today, they would be placed in that category.

The word "adolescent," while it is an ancient term with shifting meanings, is used by psychologists to speak of a period of psychic development that precedes maturity and by sociologists and by anthropologists to speak of a period between physical and social maturity. I have tried to use the word in one or the other of these two senses, as the context requires.

This difficulty in establishing terminology underlines one of the chief points of the book—that the teenager is a recent idea that may not deserve to be an eternal one.

ONE
The Teenage Mystique

America created the teenager in its own image—brash, unfinished, ebullient, idealistic, crude, energetic, innocent, greedy, changing in all sorts of unsettling ways. A messy, sometimes loutish character who is nonetheless capable of performing heroically when necessary, the teenager embodies endless potential not yet hobbled by the defeats and compromises of life. The American teenager is the noble savage in blue jeans, the future in your face.

Teenagers occupy a special place in the society. They are envied and sold to, studied and deplored. They are expected to break some rules, but there are other restrictions that apply only to them. They are at a golden moment in life—and not to be trusted.

Ours is a culture that is perpetually adolescent: always becoming but never mature, incessantly losing its none-too-evident innocence. We don't want to admit that we're grown, mature and responsible. We admire people like Ronald Reagan, James Stewart, or David Letterman, who maintain a charmingly awkward, fresh-faced teenage style into middle age and beyond. We like freshman legislators and suspect the experience of professional politicians.

We are besotted with youth—it's nature's Viagra. Teenagers are filled with new powers and the ability to use them. We respond with wonder, envy—and alarm. We know we can't keep up with these kids. We wonder if they will be able to keep their energies under control. We worry that they will run roughshod over everything that's worthwhile.

What was new about the idea of the teenager at the time the word first appeared during World War II was the assumption that all young people, regardless of their class, location, or ethnicity, should have essentially the same experience, spent with people exactly their age, in an environment defined by high school and pop culture. The teen years have become defined not as an interlude but rather as something central to life, a period of preparation and self-definition, a period of indulgence and unfocused energy. From the start, it has embodied extreme ambivalence about the people it described. Teenagers embrace the latest dances and the latest fashions. Adults fear that teenagers will go totally out of control. The teenage years have been defined as, at once, the best and freest of life and a time of near madness and despair.

Our beliefs about teenagers are deeply contradictory: They should be free to become themselves. They need many years of training and study. They know more about the future than adults do. They know hardly anything at all. They ought to know the value of a dollar. They should be protected from the world of work. They are frail, vulnerable creatures. They are children. They are sex fiends. They are the death of culture. They are the hope of us all.

We love the idea of youth, but are prone to panic about the young. The very qualities that adults find exciting and attractive about teenagers are entangled with those we find terrifying. Their energy threatens anarchy. Their physical beauty and budding sexuality menaces moral standards. Their assertion of physical and intellectual power makes their parents at once proud and painfully aware of their own mortality.

These qualities—the things we love, fear, and think we know about the basic nature of young people—constitute a teenage mystique: a seductive but damaging way of understanding young people. This mystique encourages adults to see teenagers (and young people to see themselves) not as individuals but as potential problems. Such a pessimistic view of the young can easily lead adults to feel that they are powerless to help young people make better lives for themselves. Thus, the teenage mystique can serve as an excuse for elders to neglect the coming generation and, ultimately, to see their worst fears realized.

In the first decade of the twenty-first century, America can anticipate the largest generation of teenagers in its history, one even larger than the baby boomer generation that entered its teens four decades ago. Some see these young people as barbarians at the gates, and others look forward greedily to large numbers of new consumers. But all seem to agree that having so many teenagers around will mean something important for the country. That's why this is a crucial moment to question the teenage mystique and look for more useful ways to think about the young.

I'm going to begin with a horror story, one that is not at all typical of young people's experience today. It does, however, illustrate how the teenage mystique provokes us to draw spurious generalizations from a singular abhorrent act and how it can lead to strange and destructive forms of denial.

On the night of June 6, 1997, an eighteen-year-old woman from Fork River, New Jersey, gave birth to a six-pound-six-ounce baby boy in the women's rest room of the catering hall where her high school senior prom was taking place. Her son was found dead, tied in a plastic bag in a trash can in the lavatory where he was born. His mother, meanwhile, was dancing, smiling, and to all outward appearances, enjoying what's supposed to be a magical night.

This story excited tremendous public interest, as true horrors do. Always there are questions. How could she not have known that she was pregnant? Didn't her parents, with whom she was living, know? And how about her boyfriend of two years, the presumed father? The explanation that she had taken to wearing baggy clothes didn't seem convincing.

The bigger, more fundamental question was how she could have done it. She said she believed the baby was born dead. (Prosecutors felt otherwise, and in the end, she pleaded guilty to aggravated manslaughter and was sentenced to a fifteen-year jail term.) But even a miscarriage spurs more emotion than this young woman displayed. According to one account, she touched up her makeup at the bathroom mirror after discarding her child, then emerged smiling and animated, mingling with her classmates as if absolutely nothing had happened. When faced with shocking events, people search for reasons and meanings.

In this case, an explanation was close at hand: She was a contemporary teenager, a member of a generation that's out of control. "She has come of age," wrote columnist George Will in the June 15, 1997 issue of the *Washington Post,* "in a society where condom-dispensing schools teach sex education in the modern manner, which has been well-de-scribed as 'plumbing for hedonists.' " *People* magazine used the incident as an occasion to assemble a rogues' gallery of teenagers who have been charged with committing callously violent acts. One of these was another young New Jersey woman in her teens charged, with her then-boyfriend, with killing and disposing of her newborn in a motel Dumpster in Dela-ware. New Jersey Governor Christine Todd Whitman quickly came for-ward with a $1.1 million program to cure what she called a "moral crisis" that led teenagers to kill their infants. She acknowledged that teen pregnancy was actually in decline, but added that she was alarmed at the phenomenon of teenage mothers who believe "the popular attitude that says, 'Anything goes,' including giving birth to a baby and discarding it in the trash."

Governor Whitman's statement demonstrates that the facts have far less power than what people believe is true. And what we seem to believe is that today's teenagers are uniquely threatening. One distin-guished criminologist has described a breed of lawless, heavily armed, and ruthless "teenage superpredators." There's no doubt that such people exist, particularly in some low-income city districts where drug dealing and other crimes are just about the only economic activities. But there's a temptation to see all teenagers—with the possible exception of your own children and a few of their friends—as part of this savage horde.

By giving birth to and killing her baby at the senior prom, the young woman provided neighbors and pundits with a very strong temptation to cast what happened as a parable. Many elements of the teenage mys-tique—sexuality, consumption, youth culture, hell-raising—coalesce on prom night.

Compared with most other societies, ours is short of rituals that meaningfully recognize young people's arrival at maturity. The senior prom is one of the few in which young people take an active, even enthusiastic role. It marks the end of high school, the near-universal experience of American youth, in a way that allows young people to be far more expressive than they are, capped and gowned, at graduation ceremonies. Both young people and their elders expect it to be a night to be remembered for the rest of one's life.

For older people, the senior prom conjures up gyms festooned with crepe paper and girls in frilly evening gowns. That sort of prom died most places during the 1970s. What replaced it, after a few promless years in some schools, is very different, though corsages and even cummerbunds are still involved. The event is held in a hotel ballroom or catering establishment. The girls choose drop-dead sexy dresses that make them appear as adult as possible. Transportation is often by limousine, a practice that began as a concession to parents who knew that their children drink on prom night, probably because they had. It's an expensive event. A typical prom couple spends about $1,000 all told. The contemporary prom is not a farewell to school days but a strong assertion of nearly grown-up status, a status that the society at large doesn't fully accept.

By giving birth at the prom, the young woman violated the old-fashioned meaning of the prom as a celebration of the end of a protected, almost childish mode of existence. But her act also undermined the more recent tendency by young people to use the event as an aggressive assertion of maturity. She proved herself physically capable of bearing a child, but not mentally, emotionally, or morally mature enough to handle it. She had, in a word, shown herself to be a teenager.

One element of this story that captured the imagination of those who reflected on it was the music. Not long after the young woman emerged from the women's room, she requested a song from the disc jockey. It was "Unforgiven" by Metallica, a group known for the kind of relentless, pounding sounds that give parents headaches and make them wonder what their children hear in this stuff—or indeed whether they can hear at all. "If she is like millions of other young adults," wrote Will, sounding like countless generations of elders, "she has pumped into her ears thousands of hours of the coarsening lyrics of popular music." Others found significance in the lyrics of the song, which begins: "New blood joins the Earth and quickly he's subdued."

Adults have been deploring the rawness, primitive rhythms, and carnality of young people's music ever since ragtime first became popular early in the century. It reminds them that their children are becoming openly sexual—and that they have some new moves of their own.

In fact, music is more commonly a substitute for action than a provocation. Ballroom dancing is stylized seduction, but it most often leads only to another dance. People sing the blues to tell about how miserable they are, and it makes them and their listeners feel better. Marches keep soldiers in line between battles. And heavy-metal fans—when asked how they react to the cacophonous sounds and nihilistic lyrics—tend to reply that the music calms them down. We probably can't blame Metallica. (Besides, the child in its song lives into old age.)

What did make her do it? One neighbor suggested it was the result of indulgent parenting; her mother and father gave her a car, and they even bought the gas for it. Others might blame Satan—or society. I'm reluctant to hazard such explanations, because the last thing I want to do is to seem to be making an excuse for such an evil act. But whatever the cause, the story illustrates the grotesque consequences of the teenage mystique. The young woman was unwilling to admit, even to herself, that her actions had consequence—in this case, a son. Moreover, the teenage mystique enabled those around her to deny the reality of her situation, and it allowed her to deny the gravity of her act. She accepted one of the mystique's key assumptions: What teenagers do doesn't really count.

Most of us treat the teenager as a self-evident phenomenon, an unavoidable stage of life. Adults fulminate about teenagers, children are encouraged to look forward to being teens, and those who fit the definition seem to accept it, at times reluctantly. Yet the concept of the teenager remains arbitrary and confusing.

The word "teenager" tells us only that the person described is older than twelve, younger than twenty. These seven years represent an enormous chunk of a person's life, one in which most people experience big physical, emotional, intellectual, and social changes. The word "teenager" actually masks tremendous differences in maturity between different members of the age group, and within individuals as they pass through the teen years.

Defining a person strictly in terms of age feels natural to contemporary Americans. Our society's commitment to equality seems to demand objective classifications. We don't trust people in authority to judge whether, for example, this young person is mature enough to drive or to vote, while another one the same age is not. We recognize that such judgments might be correct, but also that they are subject to abuse. Conferring and withholding rights is a serious matter, and age seems to be the most objective standard we can apply.

The trouble with creating a distinct group defined solely by age is that we conjure up phenomena that don't really exist. Is there really an epidemic of teenage pregnancy, or are women in their teens simply participating in a larger societal trend to bear children out of wedlock? Crime, especially drug crime, is a multigenerational industry in which people in their teens are active participants. Is it, then, meaningful to speak of a teenage crime problem? In 1998 Reuters reported on a scientific study that purported to show the neurological causes of "teen angst." What the researchers found was that the "teenage mind" reacts to crises while using a part of the brain associated with impulsive action, while adults make greater use of the areas associated with rationality and experience. Deep in the story, it was noted that older teens have brain responses close to those of adults. There is, in other words, no such things as "the teenage mind," only developing human minds.

Until the twentieth century, adult expectations of young people were determined not by age but by size. If a fourteen-year-old looked big and strong enough to do a man's work on a farm or in a factory or mine, most people viewed him as a man. And if a sixteen-year-old was slower to develop and couldn't perform as a man, he wasn't one. For young women, the issue was much the same. To be marriageable was the same as being ready for motherhood, which was determined by physical development, not age.

Sometimes young people could display learning, skills, or religious inspiration that would force their elders to acknowledge their maturity. The important thing, though, was that the maturity of each young person was judged individually.

Today's teenagers serve a sentence of presumed immaturity, regardless of their achievements or abilities. The prodigy has to finish high school. The strapping, well-developed young man shows his prowess not at work but on the football or soccer team. The young woman who is ready to be a mother is told to wait a decade instead.

That doesn't mean that we have given up thinking about ourselves and others in terms of size, only that this mindset coexists uncomfortably with our practice of regimented age grouping. Recent studies show that young people who view themselves as more physically developed than their peers are more likely than others to be sexually active, to drink, and to engage in risky behavior. They often cause discipline problems in schools because they are unwilling to accept society's assertion that they are not grown up. They are also more likely to attempt suicide.

Today's young people grow to their full size and reach sexual maturity sooner than did members of earlier generations. The mismatch between young people's imposing physical development and their presumed emotional, social, and intellectual immaturity is dramatic. Will these powerful young people, who are judged not yet ready to join the adult world, assert themselves and immediately career out of control, endangering themselves and others? This is a perennial anxiety that's near the heart of the teenage mystique.

Teenagers spend much of their lives dealing with people who do not know them as individuals, and under the control of institutions that strive to deal with people uniformly. Once they leave the house, they are at the mercy of a battery of bureaucracies. Chief among these are public high schools, junior high schools, and middle schools, all of which have become increasingly large and impersonal. Moreover, issues such as insurance liability and fear of sexual harassment charges have weakened relationships between students and teachers.

When the school day ends, teenagers in public are a suspect class, of particular interest to local police and the security forces of shopping malls and other private businesses. Teenagers are often expected to be transgressors, and when they do fail to conform to the frequently ambiguous rules within which they are expected to live, they can be punished very severely. Institutionally, teenagers are treated as something less than real people—sometimes resembling children, sometimes adults. And during the 1990s, it has become politically popular to punish them as both.

In recent years, adults' disapproval of teenagers has grown. In a 1997 survey, 90 percent of adults said that young people are failing to learn such values as honesty, responsibility, and respect, and two thirds agreed that the next generation will be worse than the last.

The media offer reasons for pessimism. Just about all the news they report about teenagers is bad. (Most news about anything is bad, of course. Part of the problem may be that "teen" is such a short and seemingly descriptive word for headline writers. You rarely see a headline that says FOUR TWENTYSOMETHINGS ARRESTED.)

Still, some of the stories are memorable. Young males show up at school with automatic weapons and mow down students and teachers who slighted them—or anybody who happens to be around. A teenage male murders a child who came to his door selling candy bars to raise money for his school. Two teens in a remote house order out for pizza—because they plan to kill the delivery man. A gang of teens on their way to play baseball in the park bludgeon a stranger to death with their bats. The litany can go on and on. Teenagers seem to be descending to a level of brutality beyond what many adults can remember, or even imagine.

"We know we've got about six years to turn this juvenile crime thing around, or our country is going to be living in chaos," President William J. Clinton said in 1997, expressing an opinion shared by others across the political spectrum. At the time he made the statement, he knew that crime by young people had actually been declining for two years, but he responded to public perceptions of a massive increase.

Teenagers' crimes have become deadlier and more spectacular, but that's largely because of the weapons being used. The switchblades used by the juvenile delinquents who were so menacing during the 1950s were surely lethal, but an individual wasn't able to harm more than one person at a time. When young people have access to guns, a private dispute can turn into a massacre. Indeed, many of the most dramatic recent incidents have happened in rural areas, where school violence is extremely rare, but firearms are common. A recent national study of adolescent health identified guns in the home as a measurable risk to teens' health, along with drugs, alcohol, cigarettes, and automobiles. In a 1998 survey, one in six teenagers claimed to carry guns occasionally, and 6 percent said they take them to school.

Thus, even as serious crimes by teenagers decreased by more than 11 percent between 1994 and 1996, Congress and many state legislatures considered or enacted legislation requiring people as young as eleven to be tried and punished as adults for a wide range of crimes. In the November 19, 1995 New York Times, Princeton criminologist John DiIulio provided a justification for toughness when he called the mid-1990s crime lull "the calm before the storm" and warned of a coming generation full of teenagers who are "fatherless, godless, and jobless." Today's teenagers are a menace, and tomorrow's are going to be even worse, the argument goes.

Youth crime rates have been declining throughout the 1990s. Still, as has always been the case, people in their teens—especially males— commit a lot of crimes. Throughout the 1990s, a bit less than a third of such serious offenses as murder, rape, larceny, and auto theft were committed by people in their teens. (In 1979 teens committed about half of such crimes, but that was largely because there were more teenagers then.) Violent crime for financial gain, like athletics, is a young man's activity, requiring daring, physical confidence, and to some degree, belief in one's own immortality. Such criminality tends to peak along with men's physical prowess during their twenties.

Crime is one of the few pursuits in contemporary life that allows young men to reach economic maturity at around the same time as their bodies. Before the invention of the teenager, most young men were making money on their own at fourteen or so, and they weren't considered a breed apart but simply members of the workforce. Crime is one of the few occupations to which youthful entry is not foreclosed.

We are more accustomed to thinking of contemporary teenagers as predators than as victims, but there are good reasons to worry about them. Far more of them are growing up in low-income households than was the case a few decades ago. They spend more time on their own; today's young people are able to be with their parents ten to twelve fewer hours each week than was the case three decades ago.

They are likely to attend schools that are overcrowded, a condition that will worsen because few school districts expanded their secondary schools to accommodate the larger numbers of teenagers they will enroll during the next decade. Many school districts have little choice in the matter, because they are starved for money. Public schools, new taxes, and teenagers are three of the least popular causes in contemporary America, and when you put them all together, it's a political loser.

Even those who aren't poor will have a harder time realizing their ambitions in an economy in which higher education is becoming ever more necessary and ever more expensive. Four years of college are no longer enough to bring substantial financial rewards. The median income for college graduates is equivalent to what high school graduates earned in 1970.

The good news about contemporary teenagers is that they are coping very well despite these challenges. They are, in general, far better off than their parents were at their age. They are healthier than ever, and although their risk of being shot or murdered is higher than it was forty years ago, their risk of dying during their teens is considerably lower. They are less likely to die in automobile accidents, and despite recent upticks in alcohol and drug use, these remain far below the levels of twenty-five years ago. Teenage pregnancies are in decline. So are high school dropout rates. Young people's aspirations for higher education are on the upswing for nearly all racial and ethnic groups, with the significant exception of Hispanics. They express greater optimism about their lives than young people of the 1970s and 1980s.

It's good to feel hopeful about the prospects for the young and to feel confident that today's and tomorrow's teenagers will turn out all right. Yet, sometimes this heartfelt desire to see young people turn out well gives rise to a destructive aspect of the teenage mystique. We tend to believe that young people are not fully formed and that there is still time to help them correct any mistakes they have made. This is a generous belief that contains a substantial element of truth. But this optimism becomes distorted when, seeing the teen years as the last chance to perfect troubled young people before they turn into vicious adults, our drive to perfect the young becomes coercive and arbitrary.

The belief that the teenager is an unfinished person helped give rise, a century ago, to the juvenile justice system. This placed the courts in a quasiparental role toward young people and created separate procedures and punishments—along with a large roster of offenses that are considered crimes only if young people commit them. A similar desire to shape the teenager to society's liking also underlay the early twentieth-century movement to make high school, which only a small fraction of young people then attended, into a universal experience. This ambition was achieved during the 1930s.

Although these two institutions are currently under fire for being ineffective, efforts to perfect teenagers are stronger than ever. Indeed, reforming the behavior of teenagers has become a surrogate for trying to deal with many problems of the society at large.

The weaknesses we see in youth are our own, and we know it. We become angry with teenagers because we want them to grow into healthier, wealthier, and wiser versions of ourselves. We convince ourselves that by whipping today's young people into line, society will achieve temperate perfection a few decades hence, and we will atone for our own shortcomings.

Teenagers are the target of nearly every effort to cut smoking, alcohol abuse, and illegal drug use. After all, the teen years are when most people acquire bad habits they'll have the rest of their lives. Yet, despite laws prohibiting sales of tobacco products to minors, the disappearance of cigarette vending machines, and massive advertising and educational efforts, smoking by teenagers increased during the 1990s. One has to wonder whether, by focusing so single-mindedly on teenagers, these laws and exhortations convey the message that smoking is an adult activity—not merely a stupid one. (Smoking is also an effective, low-key way to revolt against health-obsessed baby boomer parents.) Young people have for many years asserted themselves as grown-ups by acquiring adult vices.

We tend also to overlook older age groups in which drug use and drunken driving is increasing, while we pay very close attention to teenagers whose behavior has, in general, been improving. One result of this attention is a national movement to restrict the driving privileges of the young by limiting the hours they can legally drive or forbidding them to carry other young people in their car. Yet even the American Automobile Association, which supports such restrictions, concedes that the main problem with teenage drivers is not their age but their inexperience.

There has also been a widespread revival of youth curfews in cities from Washington to Los Angeles, Phoenix to Detroit, and Dallas to San Jose. By 1997, 146 of the 200 largest American cities had curfew laws requiring young people under sixteen, seventeen, or eighteen to be off the streets after a certain hour. So did many of the large suburban counties that surround the cities. In most places, the curfew is enforced intermittently and selectively. It is a tool that allows police to detain young people without cause. Meanwhile, evidence that curfews stop crime by young people is scant. Most crime by juveniles is minor and happens immediately after school gets out. In San Diego, a group challenging the city's curfew on constitutional grounds found that—during a period when the curfew wasn't being enforced—83 percent of youth crime occurred outside of curfew hours.

No part of the teenage mystique is more alluring and perplexing than sexuality. Being a teenager is, in some respects, an unnatural act, an imposition of culture on biology. It means continuing to be a child when your body is telling you otherwise. Young people nearing the peak of their physical and sexual powers are expected to delay using them, and focus these energies on acquiring skills and moral values. Adults, especially parents, hope that young people will remain innocent of their sexual power. They are embarrassed to talk with their children about sexuality, fearing that doing so will only encourage their children to have sex. But they suspect, correctly in most cases, that teenagers are already exploring their sexuality.

Teenage sexuality suffuses and confuses the culture. Like a tree in first bud, the potential adult body seems more attractive than one that is fully formed. Adults envy teens for their energy, their freshness, their passion, and they seek to imitate them. There's something crazy about the way grown-ups try to recapture an evanescent moment. Meanwhile, television, magazines, and films are saturated with seductive imagery of teenage bodies acting out adult fantasies.

That's unsettling to adults and young people alike. The worship of taut young bodies sends a message to adult women that their own maturity is a kind of failure. Young people live in an atmosphere in which erotic images of young female and male bodies are being used to sell nearly everything. (And we are only now beginning to realize how often adults' erotic feelings are acted upon with teenagers.)

Contemporary teenagers are sexually active. Among seventh and eighth graders, about one in six reports having had sexual intercourse, while among ninth through twelfth graders, nearly half report intercourse. Meanwhile, young people are changing their perceptions of various sexual acts. Some researchers have found that oral sex is increasingly defined as just another form of "making out," short of real sex, though most parents would not agree.

And although the abstinence education programs offered by many school districts advocate postponing sex until marriage, the shape of contemporary young people's lives makes that unlikely. Today's teenagers are faced with the prospect of an adolescence that stretches well into their twenties, as graduate and professional education are increasingly required for jobs paying a middle-class salary. Forty years ago, teenage marriage was commonplace; now it's close to unthinkable. It's one thing to tell a fifteen-year-old to save sex for marriage when the event is likely in three years or so, but quite different when the event is a decade or more away. Few are going to wait for those ten years, so they reason, "Why not now?"

Still, even though adults, teens, and children live in a hothouse of adolescent sexual imagery and innuendo, we persist in a belief or a hope that young people can be kept sexually innocent. We cling to a biologically naive belief that if teenagers aren't told anything about sex, the problem will go away.

Parental reticence about sexuality is nothing new. Even Sigmund Freud's son Martin complained that his father was too embarrassed to tell him anything about sex. The sex-drenched character of contemporary commercial culture would seem to demand that schools and other institutions that serve teenagers provide information that might put the media's seductiveness in context.

Nevertheless, Congress has passed a law providing extra funding to states that initiate programs encouraging abstinence as the only means of birth control. Before the 1997 school year began, parent volunteers in one rural North Carolina county gathered to slice three chapters out of the ninth-grade health textbook. The subjects considered included marriage and parenting, contraception, and AIDS and sexual behavior. Teachers were instructed to tell students that they should find out about these things at home, probably from parents who are even more squeamish about sex than Freud was.

There's no evidence that information about contraception—or even distributing condoms in school—gives young people the idea of having sex. The entire culture and their own bodies seem to be doing that quite effectively. Indeed, if there is any one thing that can make sex dull for teenagers, it is to teach it in high school.

Like sex, money plays a complex, often contradictory role in our thinking about teenagers. Teenagers' buying power is as robust as their sexuality, yet we believe that young people do not and should not play a role in the productive economy. Teenagers are to be protected from the world of work, whether they want to be or not.

Teenage consumers help drive such leading industries as popular music, movies, snack foods, casual clothing, and footwear. They spend about $100 billion a year, just on things for themselves. Two thirds of this comes from their own earnings, the rest from their parents. In addition, large numbers of people in their teens shop for food for their families and influence purchases whose estimated worth, depending on the assumptions you make, ranges from $40 billion to $100 billion a year. Because their numbers are increasing and their buying habits aren't yet fixed, teenagers are of intense interest to marketers. Their tastes and habits hint at the world of tomorrow. Teen consumers are believed to have the economic power to make a new television network succeed or to enable retailers to make money on the Internet. Indeed, because marketers find that people under twenty buy products based on their "aspirational" age, usually about five years older than their real age, young Americans become part of the "teenage" market around the age of nine.

Adults may not approve of everything teenagers do with their money, but they are complacent about teenagers' role as consumers. However, the prospect of teenagers earning money, especially a lot of money, troubles many adults.

That's because our culture tells us that the "job" of teenagers is not to work for a living, but to go to school and acquire skills that will enable them to be fully productive five or ten years later. Educators and social workers call for teenagers' pay to be kept low so that they won't be tempted to enter the job force and become independent prematurely. Underpaying young people is, thus, a virtuous act, done for their own good.

We tend to view teenagers as more or less of a leisure class, even though it's clearly not true. They are to be seen everywhere in the service economy—flipping burgers, working in stores, delivering parcels as they did a century ago, though now with a fashion-forward attitude. A recent Gallup poll found that about half of all high school juniors and seniors work part-time, averaging fifteen hours a week.

Despite the evidence of teenagers in productive positions—and the existence of entire industries that depend on their labor—we tend to view their work as inessential, a way for young people to buy the luxuries they demand while learning skills that will be useful later in life. In fact, a study by Tulane University researchers suggests that working teenagers are more likely to come from relatively affluent two-parent suburban households than from poorer, urban, or single-parent households, which isn't too surprising, considering that suburbia is where the service jobs are found.

While having a teenager in the household once gave parents useful labor and even a positive cash flow, contemporary teens are far more often a financial drain. What never seems to change, however, is the effort to harness teenagers' productivity while limiting their economic rewards and the personal independence these could provide.

Young people today seem to be in a world of their own. They go to school with their age mates. They are with them on the job. They hang out with them and they buy products and seek out entertainment designed just for them. The existence of this teenage culture, which seems wholly impervious to adult influence, is one of the most contentious aspects of the teenage mystique.

Talcott Parsons, the sociologist who was the first to study contemporary youth culture during the 1940s, concluded that it placed a higher priority on humanistic values than does the society at large. He observed that while adults are judged on a relatively narrow range of competencies, being what teens then termed "a swell guy" required a wide range of physical attributes, athletic ability, social skills, confidence, and to a lesser extent, intelligence. In practice it's very demanding to constantly undergo such all-encompassing assessments. And it's extremely inconvenient that it should come at the same time as pimples.

More recently, youth culture has taken on a more sinister aspect. Parents often feel as if their teenage offspring have suddenly become members of an alien tribe whose members pay attention only to one another. Parents feel their opinions count for nothing, compared with the judgment of the other kids at school. In reply, teenagers often complain to interviewers that they have been abandoned by their parents, who are working hard, divorced, or uninterested in them.

Those on both sides of the generational divide have a point. Some degree of withdrawal from one's parents is a necessary part of growing up, and parents inevitably find this separation emotionally wrenching. Likewise, many families have been caught between falling wages and rising material expectations, forcing longer work hours for parents. Friends help fill the gap.

Nevertheless, young people crave contact with their families. One of the most important incentives to teenage pregnancy, researchers have found, is that it is a way for the young woman to win individual attention she wouldn't ordinarily receive from family members.

Indeed, the most powerful positive factor that determines the well-being of young people, according to the 1997 adolescent health study, is the presence of parents who are engaged in their children's lives and have high expectations for them. On average, young people spend more time hanging out with people their own age. Still, just about every study that has been made of young people in their teens shows that they seek a connection with their parents and are very sensitive to their actions. The teenage mystique, which encourages parents to treat their young as if they were some strange and exotic species, plays a big role in creating the youthful anomie and deviant youth culture that adults so fear.

The upcoming generation keeps changing its shape—from tormentor to victim, from innocent to voluptuary, from consumer to creator, from menace to hope. It's not surprising; teenagers quite literally embody change. They undergo a series of physical transformations that can be discontinuous, seemingly unpredictable, and unsettling for the young and their elders alike. They become walking, back-talking metaphors for the speed and inexorability with which our lives are being transformed.

But of course they are more than metaphors, more than passive receptors and reflectors of social visions. They are individuals who will learn and create and say "No!" at seemingly inopportune times. They will, like every generation, face the difficult task of making sense of their lives in difficult times.

It's even possible that they will, with our help, revise or escape the teenage mystique and invent new and better ways to be young.

TWO
Only a Phase?

"Look, sir. Don't worry about me," I said. "I mean it.
I'll be all right. I'm just going through a phase right now.
Everybody goes through phases and all, don't they?"
"I don't know, boy. I don't know."

—J. D. SALINGER, *The Catcher in the Rye* (1951)

The idea of the teenager—and that of the adolescent on which it is based—are inventions. They emerged as useful ways of explaining and controlling youthful behavior.

Inventions like these cannot, however, materialize out of thin air. They must have some basis, not simply in history, but in the physiological, mental, and emotional developments of human beings—and perhaps also in some fundamental patterns of behavior that transcend culture. Most people assume that these inherent qualities of youth explain a great deal more than they really do. But it would be foolish to pretend that they don't exist.

In his wheedling, pathetic, smart, and screwed-up way, Salinger's Holden Caulfield, American literature's most memorable teenager, appeals to conventional wisdom to let him off the hook for behavior that he knows is stupid. Yet, as is so often the case, he's on to something important: Doesn't everybody go through phases? Aren't the teens a time when a lot of people do a lot of things that, from the safety of adulthood, they can recognize as really foolish? Isn't adolescence by its nature unstable, a time of life that requires and rewards fantasy and extreme shifts of moods and aspirations? Isn't being young a circumstance that's beyond anyone's control?

The answer to each of these questions is "Yes, but . . ." Yes, the human body goes through a series of startling changes during the teenage years, but for most people, these aren't debilitating. And, yes, our society, like all others that we know of, divides the human life span into a number of phases, but these ages of man aren't uniform across all cultures and they have changed dramatically even in our own. Yes, our society licenses experimentation by the young and usually expects and tolerates a certain amount of destructive behavior. But such testing of limits has as its ultimate purpose a realistic assessment of both one's own abilities and the nature of the world in which you live. Yes, dreams and fantasy are important in a society where your ability to create a salable self influences your fate more strongly than does your family or class. But ultimately, the society does not support overactive imaginations; it suppresses them. Yes, no one can help being young, and our society has dug a chasm between adolescence and true adulthood. But for the healthy individual, a sense of self that continues from childhood into adulthood manages to overcome such barriers.

Holden Caulfield's furious, funny voice is so unforgettable that his character has helped shape how several generations think about teenagers. His search for integrity and his aspiration to kindness make him a sort of hero. His behavior, however, is self-destructive and uncontrolled, and he feels miserable. Nevertheless, when the book came out in 1951, psychologists generally believed that psychological upheaval similar to what Holden experiences in the novel was a normal, even necessary part of adolescence.

Over the last four decades, Daniel Offer and others have demonstrated otherwise. Offer's research indicates that about 15 percent of teenagers have serious psychological problems: a significant portion. But it is no higher than the number of adults with such problems. Teenagers are just like people, in other words. They face crises peculiar to their time in life, just as adults do. Still, it's not normal to be overwhelmed by them. People do work their way into and out of serious problems in adolescence. They cannot, however, experience the phase passively, like a rider on an elevator. They have to take an active role in changing their lives. The belief that change will simply come is a dangerous one if it discourages young people, their parents, and others from coming to terms with a problem.

Offer's work, which relies on survey research, has been replicated internationally. The sense adolescents have of themselves varies somewhat from culture to culture. Still, adolescents' basic confidence in their own abilities, their futures, and the importance they ascribe to their parents appears to be universal.

The overwhelming majority of young people find a way to live more or less successfully in the society of which they are a part. Human beings are adaptable.

Yes, we knew that. But Holden's question about how people go through phases seems to require a deeper answer, one that reveals something profound about human nature. We like to assume that there can be a kind of baseline so that we can determine what aspects of teenagers are inherent, which are cultural, and which are responses to historical circumstance. The rest of this chapter and the one that follows are about searching for this baseline.

Most would begin this quest to determine which aspects of teenage life are inherent and universal with "hard" science. Chemistry appears to be more fundamental than physiology, which is in turn prerequisite to mental development. Social relationships and issues of power and property seem to be the least decisive in determining the nature of the organism.

This bias is expressed in the two-word term that serves as the vernacular explanation for almost everything teenagers do: Raging Hormones. Even though nobody has identified a heavy-metal hormone, a speeding hormone, or even a getting-laid hormone, we ascribe everything loud, fast, or upsetting that teenagers do to their endocrine secretions.

Even women who consider themselves feminists sometimes fall back on the raging hormone explanation for teenage behavior. Since hormones once served as an excuse to find women incapable of men's work, it's worth asking whether a similar kind of thinking is also doing a disservice to teenagers.

Perhaps the most surprising thing about teenagers' raging hormones is how little they have been studied. There have been a number of important recent studies about the mechanisms that trigger and control the physical and behavioral changes of puberty. Still, scientists probably have a more systematic understanding of the first few minutes of the universe than they do about the several years that produce the physically adult person or person.

We do know that this is a long process. It starts well before young people enter their teens and continues into their twenties. In fact, one of the most concentrated periods of hormonal activity, especially for males, happens after the teens when they undergo their final muscular bulking-out. Sexual stirrings are evident by age nine or ten and breast development begins in most American girls shortly thereafter.

Different hormones closely linked to the development of sexual characteristics come into greater play as young people approach their teens. The growth spurt, the development of primary and secondary sexual characteristics, the change in body shape and musculature, and the voice change in boys all offer indirect evidence of hormonal activity. Yet, at the ages of fourteen and fifteen, the years in which the hormonal storm is popularly assumed to be raging most violently, especially in males, hormonal activity is actually less intense than it is at more outwardly peaceful times.

The discontinuity of the process—the way some things seem to be happening instantaneously, others awkwardly, and still others embarrassingly late—can make it seem out of control. A boy of thirteen is growing faster than he did since he was putting his parents through the "terrible twos." All that growth sometimes makes him very, very tired, a physical response that parents sometimes mistake for an emotional one. Girls move through the transition sooner than boys, something that's evident in junior high schools. The overall sequence, while its timing might vary greatly between individuals, is predictable detail by detail. A boy can be fairly sure, for example, that when his voice begins to crack, his penis has reached its full size (though he can still hope).

In summarizing so quickly, I am leaving out some nuances, and perhaps implying more certainty and consensus than really exists. Still, what our knowledge to date suggests is an exquisitely orchestrated process—one whose result is a body in flux but not a person out of control. There is little doubt that teenagers, like adults, think a lot about sex. The development of genitalia naturally turns the mind toward their possible use. But while endocrinology is a field where fundamental discoveries are made regularly, there is not yet any biochemical explanation for surliness, self-absorption, or rebelliousness.

This failure to find biological explanations for cultural traits undermines a certain amount of pop wisdom, but it doesn't really upset the hierarchy discussed earlier. What does throw such thinking into question is the clear evidence that historical and cultural forces have a very strong impact on teenagers' physical development.

At the end of the twentieth century, an American female typically first menstruates at the age of twelve years and nine months. A century ago, this milestone was reached about two and a half years later, at more than fifteen years of age. It's difficult to know what the norm was in the middle of the eighteenth century, but it was likely a year or more later. This meant that a woman's sexual maturity often coincided with her reaching marriageable age. Very likely, our society has reached some sort of record in the length of the gap, close to a decade, between the time a middle-class girl reaches sexual maturity and the time when it's socially respectable for her to marry.

This earlier sexual maturity is probably the result of improving nutrition. There is evidence that the abundance of food in North America started accelerating physical maturation well before the same phenomenon was seen in Europe. This historical finding is consistent with a recently published hypothesis that female puberty is triggered when a girl reaches a critical amount of body fat and produces a hormone called leptin.

Everything—even teenagers' bodies—varies with time and place. The fifteen-year-old girl we find on an eighteenth-century farm was at a different stage of physical, and very likely psychosocial, development than the fifteen-year-old girl we find working in a mill a century later, who was in turn different from the high school sophomore of 1999.

In the past, such changes in the biology of the teenager were inadvertent by-products of economic or social phenomena. Now, however, it's possible to intervene to a limited degree in such development. Genetically engineered human growth hormone has been available for more than a decade. It is prescribed not only for people with serious hormonal deficiencies, but also for those at the short end of the normal range of height (though clinical trials have cast doubt on the effectiveness of such use).

Children in the lowest 3 percent, or by some definitions 5 percent, of the range of heights are now defined as having normal variant short stature, abbreviated SS. Although the term contains the word "normal," the short form SS gives it the aura of something that really ought to be treated medically. Parents' motives for doing so range from making it possible for their children to realize athletic ambitions to sparing their children epithets like "shrimp," "squirt," and "peewee." (The shortest boy in my elementary and high school classes spent close to a decade answering to the name "Weaselgrunt," which must have had some psychic impact.)

This is obviously more than a purely cosmetic matter. There is plenty of evidence that teachers and others in authority treat students of shorter than average height as younger than their age, and thus do not challenge them intellectually or take them seriously. The human growth hormone thus serves as a biochemical solution to a social problem. All parents want their children to be above average.

While chemical aids for adapting to society aren't new, the artificial growth hormone is highly specific and powerful, and a portent of more to come. Indeed, it's probably possible to synthesize all the elements of the hormonal storm we assume to be raging in the young, and to use these to fine-tune at least the endocrinological development of the adolescent.

Let's pretend that's possible right now. How would we redesign our teenagers? Would we want the growth spurt and physical maturity to come later than it does now? Would we look for a way to postpone females' ability to conceive? Can humans achieve the intellectual and moral breakthroughs that are characteristic of these years without the bodies that come with them? And if we did, who would play football? Who would go to war?

This line of speculation is scary. We don't know enough about how hormones work. There is, however, reason to believe that they help shape behavior and mental capacity as well as physical development. What makes it even scarier is that it's easy to imagine someone trying to do this for real. Finessing the social or moral problem by finding a technological solution for it is, after all, the American way.

Synthesized hormones will no doubt play a role in the future, but the reality of the teen years is far more than a physiological storm. It is also a matter of psychic development, and most of all, it is a set of cultural expectations. If you expunge the biochemical determinism from the questions above, you come up with issues that societies have grappled with since ancient times. What does it mean to be grown up, and how should it be recognized? What is important for young people to know, and how should they be taught? What special role should young people play in society? How do children contribute to the wealth of their parents, and how should that wealth be allocated to the children?

These are issues that are implicit in the project of contemporary, technologically advanced society to classify physically developed people as socially immature for a steadily increasing portion of their lives. The institutions and expectations we have invented to deal with such questions are complex, and they interact in ways we do not fully understand. Teens are buffeted by discordant messages from home, from school, from the media, and from their own experience that must be at least as disorienting as those hormones we imagine to be raging within.

When Holden Caulfield talks about the phases people go through, though, he is not talking about either the actions of his glands or the strictures of his society. He's talking about psychological changes, storm and stress brought on not so much by raging hormones as by the challenges inherent in his time of life. In the United States in the immediate post–World War II era, people sought to understand adolescence mostly in the psychoanalytic terms proposed by Sigmund Freud. Adolescent psychology, a field invented by the American educator G. S. Hall at the turn of the century, rested on Freudian ideas and also on Darwinian evolutionary theory and a general faith in progress.

The word "adolescent," which comes from a Latin root meaning "to nourish," derives from a word used by the ancient Romans and came into English, via French, in the fourteenth century. It meant someone who was still growing, often but not necessarily a teenager. Since the turn of the twentieth century, however, adolescence has referred to a specific period of life fraught with a series of difficult psychic challenges. Generally, adolescence has been assumed to begin at puberty, although the Carnegie Commission on Adolescent Development has recently sought to push recognition of its onset back to ten years of age.

One of the key premises of Hall and his colleagues was that the adolescent might suffer from symptoms that would be considered mad in an adult, but are just the part of normal mental development for the young person. Anna Freud believed that the reason we don't remember our adolescent years is that they were so troubling that we suppress them. "What we fail to recover, as a rule," she wrote of adult memories of youth, "is the atmosphere in which the adolescent lives, his anxieties, the height of elation or depth of despair, the burning—or at times sterile—intellectual and philosophical preoccupations, the yearning for freedom, the sense of loneliness, the feeling of oppression by parents, the impotent rages or active hates directed against the adult world, the erotic crushes—whether homosexually or heterosexually directed—the suicidal fantasies, etc."

The rise of adolescent psychology was one of the key events in the history of the American teenager. It established the adolescent as a special, unstable sort of creature. Moreover, it gave rise, soon after Hall's *Adolescence* appeared in 1904, to what might be termed the Holden Caulfield excuse: "I'm only an adolescent, so I'm not responsible for what I do."

The place for a detailed discussion of Hall's ideas and their implications is in its historical context. What is important at this point is to note the way that the idea of the adolescent seems to deny either historical or cultural context. The adolescent was seen as someone who always had been and always will be.

Adolescence as a painful, stormy, yet precious and crucial passage in one's life was not a wholly new idea. Its greatest expression is probably *Émile*, Jean-Jacques Rousseau's description of the ideal education of one boy. The premise of this philosophical narrative is that Rousseau has undertaken to raise an imaginary, noble boy literally from the cradle until he is ready to take his place as a good and civilized man. Because he believes that society leads young people astray, he has raised the boy in isolation. But despite his being sheltered from corrupting conditions, at fifteen or sixteen, Rousseau writes, a crisis is on the horizon:

> As the roaring of the sea preceded a tempest from afar, this stormy revolution is proclaimed by the murmur of the nascent passions. A mute fermentation warns of danger's approach. A change in humor, frequent anger, a mind in constant agitation makes the child almost unmanageable. He becomes deaf to the voice which made him docile. His feverishness turns him into a lion. He disregards his guide; he no longer wishes to be governed. . . . This is the second birth of which I have spoken. It is now that man is truly born to life and now that nothing human is foreign to him.

This description, while more eloquent than most, is certainly a familiar recounting of what, even today, are viewed as the chief symptoms of adolescence. Rousseau makes it clear that what has happened to Émile is a result of the emergence of sex drives.

Rousseau's recommended approach to raising the young man was "to delay the progress of nature to the advantage of reason." His teacher even sleeps in the same room as his student to make sure that he never masturbates. Rousseau says the teacher must seize this moment to channel the boy's maturing desires toward learning. "This age never lasts long enough for the use that ought to be made of it," Rousseau writes. "That's why I insist on prolonging it. Make progress by sure steps. Prevent the adolescent's becoming a man until the moment when nothing remains for him to do to become one."

Toward the end of adolescence, Rousseau contrives for his student to happen upon a wise and beautiful young woman: Sophie. He is ready to meet her because he's able to appreciate both her beauty and her wisdom. Even in this case, there must be a delay of consummation, as Émile is sent to travel and live in the society of which he and Sophie will be exemplary members. The end of Émile's education, the moment when he is no longer a child in any respect, is when he and Sophie marry and are ready to have children of their own.

Émile is one of those famous books that hardly anybody reads. It is often described, simplistically, as an argument for mankind's essential goodness. Nevertheless, it permeates our thought and institutions. Its identification of eros and intellectual achievement, and its prescription to delay and channel the eros to force maturation in other directions under-lie our educational institutions, and indeed our concept of adolescence.

Rousseau wouldn't buy the hypothetical maturity-delaying hormone treatment I imagined above, because it would hinder young people's intellectual and moral development as well. His assumption is that there are gratifications other than sexual ones that require the energies, emotions, and attachments that accompany erotic awakening. Sex, in a sense, is easy. The parts are all in place. Other sources of lifelong satisfaction are more difficult. They depend on hard work and preparation, and thus on the delay of easy gratification. (One who held the opposite view was Bertrand Russell, who believed that young people should get their first sexual encounter out of the way so that they would stop obsessing about it and study their math.)

Adolescent psychology incorporated Rousseau's stormy revolution of the spirit, his belief that maturity comes not with the onset of erotic impulses but their control and, finally, their consummation in a hetero-sexual relationship. But while *Émile* is a thought experiment, in which only Rousseau the teacher and nature itself shape the young man's life, adolescent psychology claimed to be a science.

Moreover, the adolescent psychology movement added a crucial, though severely complicating element to the story—the family. The changes that come to the young person's body are thought to reopen the unresolved struggles of infancy.

In this view, what the thirteen-year-old and the two-year-old have in common are physical growth spurts, and far more important, a need to separate from their parents that is as painful as it is necessary.

The big difference is that the thirteen-year-old is so much more capable than the two-year-old, whose chief powers reside in crying and saying no. If properly directed, young people can use their imaginations, their emotional expressiveness, their self-control, and their capacity for learning and understanding to navigate the psychic minefield of adolescence. But their greater capacities can also be directed against their parents, themselves, and their own growth. There is danger both in resolving things too quickly, which is a way of not really facing things, and in being so overwhelmed that you are unable to make use of your capacities.

Hall added another element to this story. This was the idea of recapitulation, the belief that the development of the individual mirrors the evolution of the species as a whole. Thus, he saw the adolescent as a savage, prone to violent, disruptive, impulsive behavior. The good news in this was that, just as humanity evolved to a higher form, adolescents will grow out of their savagery. If your teenager is a Neanderthal, you can take heart. It's only a phase he's going through. While nobody takes recapitulation seriously anymore, the optimism inherent in the notion that adolescence is something you'll eventually grow out of does survive. Indeed, it's probably a stronger presence in most people's minds than the more central concept of resolving family conflicts and achieving autonomy.

The hope of evolution, if not recapitulation, is certainly part of the appeal of the work of the Swiss child psychologist Jean Piaget, whose approach seems closer to Rousseau than to Freud. Piaget observed distinct stages in the intellectual development of the child. He saw the years from twelve to fifteen as the final stage of transition from a concrete way of thinking to one that is comfortable with abstractions. That brings with it an ability to see oneself objectively within a larger context. It also means an ability to think about complex hypothetical situations and assess the consequences of potential courses of action.

The evolution that Piaget traces from infancy through adolescence is the development by the child of several different and distinct models of life. Piaget's child is not a quivering mass of hungers and insecurities, but rather a scientist, developing hypotheses, then trying them out.

Piaget based his theories on the close observation of a very small group of children, and he has been criticized for both class and cultural bias. Still, survey data seem to bolster Piaget's assumption that the child's drive, particularly in adolescence, is toward developing a realistic picture of the world.

Piaget's work offers alternative interpretations of what have been termed classic adolescent behaviors. For example, intellectualization—the tendency by older teens to become obsessed with logical proofs of God's existence, or hunt for hidden meanings in rock lyrics—has traditionally been seen as a way of avoiding important issues. Piaget's analysis suggests that it might simply be a matter of trying out one's newfound mental capabilities.

One of the most important claims of adolescent psychology, at least during the first half of the century, was that the phenomena it describes are universal. It said that in every place and at every time, humans have suffered the insult of expulsion from the womb. It then follows that they must find ways of defining themselves and controlling their feelings. Ultimately, in the struggles and suffering of adolescence there comes a crisis and, if things go right, its resolution.

The discovery and description of the adolescent as a universal concept spurred an interest in young people in different places and times. The goal was not to see how young people lived their lives at different times, but rather to prove that adolescents had always existed in much the same way—whether in the waning days of Hapsburg Vienna, the Athens of Aristotle, or the Hippo of Augustine.

If you look, it's possible to find wonderful passages that seem contemporary, and even timeless. An eleventh-century Japanese girl complains of her mother's "extremely antiquated mind." Aristotle's description of youth as "prone to desire and ready to carry out any desire they may have formed into action" sounds a familiar note. He added, "They are changeable too and fickle in their desires, which are as transitory as they are vehement." Besides, said Aristotle, they're always thinking about sex.

These quotations are from an influential 1964 anthology, *The Universal Experience of Adolescence* by Norman Kiell. This tome consists of passages from autobiographical writing throughout history that demonstrate aspects of the adolescent psyche, particularly as Anna Freud described them. As an example of risk-taking behavior, for example, there is a passage in which an eighteenth-century Italian describes how gangs of boys would steal horses and rampage through the countryside. These joyriders were not only in the textbooks; their counterparts were in the newspaper every day and in movies like *Rebel Without a Cause*. Except for a few details, this recollection by Montaigne could be a posting on the Internet: "I was always ready to imitate the negligent garb still to be seen among our young men—my cloak across the shoulder, my hood to one side, and a stocking in disorder, all of which was meant to show a proud disdain for these exotic trumperies and a contempt for everything artificial."

These and the hundreds of other passages in this and similar books prove that people have had many of the same feelings over time, and that people keep getting themselves in the same awkward situations. Youthful impetuosity, bravado, and awkwardness also appear throughout many literatures. These probably are common elements of being young. But that doesn't mean that the storm-tossed adolescent is an inevitable and universal phenomenon.

When Holden Caulfield talks about the phases people go through, one imagines a note of hope in his voice. Soon, no doubt, all the misery he's going through will be over. Had he actually studied the psychoanalytically oriented writings that dominated thinking about adolescence in his time, he would have been much less hopeful.

Holden seems to have been nourishing what the influential psychoanalyst Peter Blos called "the rescue fantasy"—the hope that the environment will change and what seem to be problems now will disappear without effort. Blos says it's dangerous to indulge in that fantasy. Each challenge of coping with Mom and Dad still remains. Each failure can lead to neuroses in later life, symptoms of "incomplete adolescence." Worst of all is "miscarried adolescence," which leaves a person completely unable to cope with being an adult.

While most of the psychoanalytic literature on adolescence relies on studies of severely troubled young people, its concept of normality is often synonymous with absolute perfection. In particular, its concentration on what has come to be known as the nuclear family seems to doom the many children who grow up in other sorts of environments. As one reads of the crucial developmental challenges children face and the role of both parents as sources of ideals, eros, conflict, and approval, the fatherless or motherless child does not seem to have a chance of growing up satisfactorily.

Yet, in most times and places, disease, hunger, and the perils of childbirth have made the family of mother, father, and children more the exception than the rule. Psychoanalysis is a product of the late-nineteenth-century view of the family as bourgeois refuge, and, in many respects, it is also a description of the problems that ideal brings into being.

Despite their claims to universality, the much-watered-down psychoanalytic views that underlie popular discourse on the problems of youth are time-bound and culture-bound. They described a very narrow bourgeois milieu at the time they were conceived, and they triumphed in America at the time that middle-class comforts were becoming available to everyone. But many things have changed, including our ideas of the roles of each sex. The nature of adolescence must inevitably change as well.

Erik Erikson, inventor of the term "identity crisis," is perhaps the most lastingly influential of the psychoanalytic thinkers on adolescence. One of his most important contributions was to find a place for history in the forging of identity, which he viewed as the crucial task of adolescence. "Adolescence, then, is a stage in which the individual is much closer to the historical day than he is at earlier stages of child development," Erikson wrote. "While the infantile antecedents of identity are more unconscious and change very slowly, if at all, the identity problem itself changes with the historical period: this is, in fact, its job."

Erikson, like his psychoanalyst colleagues, placed adolescence in a life course continuum that begins with separation from the mother and is resolved by commitment to a heterosexual relationship. But he also recognized that in order to marry and have a family, you must have money and skills. Worry about how you are going to make a living is not a diversion from problems within the psyche. It is, rather, one of the ways in which the self is expressed in the world.

Erikson defined adolescence as a period of moratorium, a time for young people to integrate their skills, their knowledge of themselves, the judgment of their contemporaries and their elders. What should result is a feeling of wholeness and consistency, a sense of continuity between what you have come to be in childhood and what you expect to be in the future. A moratorium sounds like a peaceful thing, a far cry from Rousseau's flash flood or Hall's storm and stress. Erikson did say that some youths, particularly those who are gifted and who identify with coming technological trends, will probably make it through the period quite peacefully.

But, he added, what is at stake is serious, so if the process is thwarted, it has consequences. He writes: "Should a young person feel that the environment tries to deprive him too radically of all the forms of expression which permit him to develop and integrate the next step, he may resist with the wild strength encountered in animals who are suddenly forced to defend their lives. For indeed, in the social jungle of human existence there is no feeling of being alive without a sense of identity."

It is possible to imagine societies that are so static that such components of identity as occupation, technical skills, and one's place in the community could pass from generation to generation unchanged. In such cases, Erikson said, the passage of generations can be ceremonialized in a way that is meaningful to all in the community. But Erikson argued that modern society was changing too rapidly for the older generations to create meaningful institutions to symbolize maturity, let alone pass along traditions they had inherited from their forebears. Indeed, he added, it was changing so rapidly that the young are unable to make any effective resistance to the change. They'll only hurt themselves trying.

While it was not the principal focus of Erikson's writings, his view of the historical dimension of the adolescent moratorium is powerful. It's natural that the expectations elders have for youth are going to be vague. Those old enough to be parents of teenagers know that they cannot predict with any confidence the world in which their children will live. Teenagers have to spend a lot of time talking with and exploring with their peers. Doing so is one of the best ways they have for predicting their own futures. The young "take things for granted" that were major struggles in the lives of their parents. Although this seemingly casual ingratitude often disturbs parents, young people have struggles of their own. They can't afford to waste energy reliving the crises of their parents. The adolescent moratorium is the time to get a highly realistic picture of the world as it is.

The one element of this analysis that seems dated is that the fast-changing society Erikson described has accelerated still more. He assumed that it's possible to define during adolescence an occupational identity that can be maintained throughout the rest of one's life. That appears to be less and less likely, as technological change and corporate restructuring force people well into middle age to reassess aspects of identity they thought they had determined decades before. Today's adolescents may even have to deal with parents who are going through some of the same crises they are coping with themselves.

Today's youth will not be able to use occupational identity as a foundation stone of their sense of themselves. They will have to find substitutes that are more durable while permitting greater change. Older people may complain about the hardships they had to endure when they were growing up, but when it comes to the things that really count, it never gets any easier to be young.

Erikson once observed that the freedom to change—the belief that one can suddenly become someone else—is a particularly American dimension of identity. There has probably never been a culture in which the quest for an individual identity has been as important, and thus as fraught with problems, as that of the United States. Americans believe that each individual is unique. We learn that any boy or girl in the country can grow up to be a billionaire, a rock idol, or president of the United States. This individualist myth sometimes, it is true, produces a good deal of conformity, as many people seem to mold themselves into the same memorable individual. Advertising aimed at the young often urges them to express their inner selves by purchasing a product. The willingness of young consumers to buy mass-produced affirmations of their uniqueness may belie the notion of a nation of individuals. Still, such advertising could not be successful unless it was speaking to a pressure felt by members of its target market.

In our mythology, each person must find his or her own role in life. We all know that some people are born into circumstances that give them considerably more freedom than others. In returning to first principles, we return to the rhetoric of the Declaration of Independence, with its self-evident truths and Creator-bestowed rights. Rarely do we think that people are, or ought to be, born into a particular role in life, though that has probably been the norm in most times and places. Nor do we view a role in society as being a gift that its holder can pass on to a protégé. We expect the younger person to have to compete fairly for the role, through a process that is, somehow, impersonal. Indeed, we assume the elder incumbent probably doesn't fully understand what ought to be required of a successor who will face a whole new set of challenges.

There is no question that having an influential family or being known to the right people can benefit an American's career. But we tend to view this as an example of life's unfairness, not as evidence that a natural order prevails. Often, it seems that being known as the child of an important member of the community—the minister's son, for instance—

poses an impediment to identity formation. People don't come out of nowhere, but we pretend we do.

The belief in self-creation is, very likely, one of the glories of American culture. It seems likely that in cultures where more of one's identity is preordained and each individual need not start from scratch, some of the trials we associate with adolescence would disappear. Is that true? Do other cultures magnify other aspects of the maturation process? Are there some aspects of adolescence that we choose not to recognize? One way to approach these questions is to take a look at the way some other cultures think about what we assume to be problems inherent in adolescence.

THREE
Coming of Age in Utter Confusion

> In the region [of Central Africa] where feminine beauty is all but identified with obesity, the girl at puberty is segregated, sometimes for years, fed with sweet and fatty foods, allowed no activity and her body rubbed assiduously with oils. She is taught during this time her future duties and her seclusion ends with a parade of her corpulence that is followed by her marriage to her proud bridegroom.
>
> —RUTH BENEDICT, *Patterns of Culture* (1934)

Other cultures often look crazy from the outside. Benedict's description of the "fatting house," in which young women in their teens were treated like Strasbourg geese, seems bizarre and grotesque.

Here's another scenario that may be even stranger. Throughout her teenage years, a young woman is bombarded with advertising for fatty snack foods, and she is constantly being asked, "Fries with that?" At the same time, she is told constantly that her future happiness depends on being extremely thin. This is the situation young women face in contemporary America. It's an odd practice, but we tend not to think it so. It's how we live.

The Central African practice, the expression of a "primitive" culture, had at least the virtue of being internally consistent. It prepared the young woman for the next stage of her life. The contemporary American practice has failure built into it. For the last twenty years or so, eating disorders have become epidemic among young women (and they have become more common among young men as well). Obesity is also increasing among people in their teens. Living in a junk food culture, their minds marinated in unrealistic images of what their bodies should be, contemporary teenagers must find a way to balance these powerful contradictions and develop a healthy sense of themselves.

Such challenges are not inherent in either adolescent physiology or psychology. Rather, they are endemic to contemporary American culture. From inside the culture, such dilemmas come to seem inevitable. A culture is, after all, little more than a group of beliefs and practices that its members accept without very much thought. Those who fattened their daughters saw it as a step to marriage. We see adolescent confusion as an inescapable part of growing up.

We are, at the moment, especially prone to see our own beliefs about the young as universal because we are exporting our culture so aggressively. Teenager-like creatures are emerging in Kuala Lumpur, Bombay, Nairobi, and anywhere MTV and other media of mass persuasion are beamed from the sky. Remote "peoples," untouched by the culture that creates teenagers, are far rarer now that they were early in the century when Benedict and other pioneering anthropologists were doing their work. (Because I'm talking largely about customs that were recorded before modernization became pervasive, I am using the past tense when describing them.)

Thus, examples like the one cited above, though distant in time and place, can be useful for thinking about contemporary teenagers. They illustrate the diversity of thinking among different peoples about what maturity is and how it should be recognized. Strange stories from exotic locales can provide perspectives on our own culture and its weird and painful practices.

The ways in which societies prepare young people to become adults and the rituals through which they are initiated express their cultures' view of what it means to be a person—both as a member of the community and as an individual. Ceremonies are an opportunity to enact and reaffirm core values, and the acceptance of the young into full membership in the community is usually a significant event.

By contrast, becoming an adult is a highly ambiguous event in our own culture. We have many ceremonies of limited significance, such as religious confirmation, high school graduation and the senior prom, or going away to college—all of which seem to promise entry into the mainstream community, but actually lead to further periods of immaturity. These are accompanied by a welter of laws that confer adult privileges and responsibilities at various ages.

Some stepping-stones of maturity are established within families: old enough to be left home alone, old enough to pierce your ears (or whatever), old enough to date. Many other thresholds are set by society in the form of laws: old enough to drive, old enough to be a soldier, old enough to vote. And there are some gray areas where the official threshold seems out of synch with practice: old enough to drink, old enough to work, old enough to consent to sex.

Over the last 350 years in America, all of these thresholds have been moved back and forth many times as we have struggled with when and how the society should accept the young as full members. We seemed, during the 1960s and 1970s, to be moving toward standardizing legal adulthood at eighteen. Since then, however, some age restrictions have been extended upward to twenty-one, while the age which one can be tried and punished as an adult has been getting steadily younger. And curfews for young people, which had largely disappeared, have been revived.

Benedict argued that coming-of-age ceremonies, where they exist, enact what it means to be an adult in that culture. Perhaps because most American adults see themselves as still young, we are particularly reluctant to acknowledge the maturity of our children.

An ironic result is that age limits established to keep young people from endangering themselves—such as minimum ages for drinking liquor, smoking cigarettes, and gambling legally—become important passages to maturity. The mark of adulthood in America is the license to indulge in bad habits.

While we have no way of telling when a young person is truly grown up, there are a couple of things we are pretty sure we know about teenagers. The conventional wisdom has it that we can expect trouble: Being a teenager is an ordeal, and being the parent of a teenager is even worse. Such wisdom also has it that teenagers are a distinct group unto themselves, no longer children yet not adult, whose time is best spent in institutions and settings populated almost exclusively by people their own age.

Ours is not the only society that has acted on these beliefs. Indeed, the segregation of young people from the rest of the society was observed by researchers in Africa even before it became commonplace in the United States. Still, neither of these tenets of conventional wisdom is true for all, or even most, cultures. The diversity of human behavior is amazing. What many societies suppress—sex between boys and teenage males, bestiality, premarital pregnancy—others celebrate.

Woman's maturity is frequently tied to her body, while man's is defined by the role he will play in society. While menstruation is an involuntary event that often triggers initiation rituals for females that lead quickly to marriage, there is latitude to judge the maturity of a male. One North American tribe placed the triggering point for a boy's initiation as the first time he dreamed of an arrow, a boat, or a woman. Because this event was unverifiable, the young man's family could manipulate it to suit their convenience.

In nearly all the societies we know about, people pass through distinct times of life. But the timing of these transitions, the activities that mark them, and the meaning attributed to them differ dramatically from one culture to another. In some tribes, the initiation into adulthood comes at ten years old or even younger, while others demand that young people be nearly twice as old. Some move directly from childhood to full adulthood with only a few weeks or months of intermediate steps, while for others, the transition can last more than a decade. Indeed, this very variation demonstrates the maturity is first and foremost a social phenomenon and only secondarily a biological one.

One anthropological term that has become part of our vocabulary for talking about the young is "rites of passage." This phenomenon was first named and described by Arnold van Gennep in a work published in French in 1908. For van Gennep, adolescent initiation was only one of many rites of passage found in numerous societies, all of which follow a three-part scheme. First, individuals are separated from their everyday life and surroundings. Then they undergo some period of transition that can last minutes or years. Finally, they are reintroduced to normal life, but their role is new and others look upon them differently.

The middle phase, that of being neither here nor there, is the most provocative and the most mysterious. You have left one realm, but you have not yet entered another. Anthropologists call this state "liminality," a word derived from the Latin word for threshold. In some initiation rites, the threshold is literal—a gate to be walked through, a portal to be entered.

A person standing at such a threshold must be invited to come through the door. But those with the power to extend the invitation are often ambivalent about surrendering their authority over the young people at the gate and welcoming them into the circle of adults. The young people, for their part, wait at the threshold because they fundamentally accept the values of the society of which they expect to become full members. But sometimes they are kept waiting at this door for years, and they must force elders to finally accept their claim to adulthood.

The inability to classify someone in this in-between state implies a kind of social invisibility. It might sound like a nightmare right out of Kafka. Nevertheless, this state of liminality can be a uniquely privileged moment in one's life, one in which behavior denied to either children or adults is tolerated, or even expected. Because the young people are not quite visible, and certainly not fathomable, adults avert their eyes from what they do. Theft, rowdy and violent behavior, and sexual activity that would ordinarily not be tolerated are condoned. Many initiation rituals incorporate cross-dressing between the sexes, a final moment of ambiguity before sex roles become fixed.

In our own society, liminality is what makes much of youth culture possible. For example, parents tell their young children how to dress and set standards for one another. But they understand that they can't control the dress of their teenage offspring, who are free to adopt transgressive, provocative costumes they won't be allowed at any other time of life. Similarly, because adults assume that they cannot understand teenagers' music or humor, most simply don't pay any attention. They thus abdicate entire realms of expression to young people saying rude, though not very mysterious, things. Young people seem to rule popular culture worldwide, yet they get much of that power, paradoxically, because the majority of people ignore those media they are perceived to dominate.

When we say, "I just don't understand teenagers," we are echoing what adults have said in most times and places about people who were no longer children but not yet adults. The same is true when we tolerate rowdy or reckless behavior from teens that would be more likely to be punished if a child or adult did it. And the fear and panic that periodically overtake society's thinking about the young must contain some echo of the emotion people felt when they banished their young, either symbolically or literally, to the wilderness. The desire by families and villages to see their young safely through this threatening and perilous state often requires huge expenditures of wealth. That's something that parents paying their children's college bills can relate to.

In many times and places, the transition from childhood to adulthood has meant literal banishment from the family, whether into the wilderness, another household, or perhaps boarding school. Our own most common ritual of physical separation—sending the kids off to college—

comes later in life than is typical in most societies. But, like banishment practices in other societies, those who participate in it, both old and young, feel that it is necessary. And both generations believe that when young people return from the banishment, they will be changed in ways that allow parents and children to view each other as fellow adults.

Several American Indian tribes followed a practice of voluntary banishment called the vision quest. Unlike most initiation rites, this was entirely solitary and self-administered. The young man would leave the group, go to a distant and perhaps sacred place, fast for several days, and wait for dreams or visions that would affirm that a guardian spirit had entered his life. Then he would return to his people, a new man, guided no longer by his family but by the spirit he had found on his quest.

The vision quest has great appeal to contemporary North Americans, not simply because it happened here, but because it seems a precursor to our individualistic culture. Moreover, it is associated with the kind of deep learning that parents know they cannot provide, but which our culture assumes will be provided by higher education. We also count on schooling to provide skills for earning a living that are infinitely more complicated than those needed by the nomadic participant in a vision quest.

The great appeal of the vision quest lies in its simplicity and brevity. Coming of age in contemporary North America is, by contrast, a procedure of mind-boggling complexity and seemingly endless duration. The banishment involved is sometimes virtual, sometimes literal, but always ambiguous.

We share the conviction of many tribal societies that the years immediately after puberty are a very dangerous age, and young women's sexuality is particularly threatening. This is true, though we do not go to the lengths that Benedict described among the Carrier tribe in British Columbia early in the twentieth century: "Her three or four years of seclusion was called 'the burying alive' and she lived for all that time alone in the wilderness, in a hut of branches far from all beaten trails. She was a threat to any person who might so much as catch a glimpse of her, and her mere footstep defiled a path or a river. . . . She was herself in danger and she was a source of danger to everyone else."

One senses in that story both a relatively common fear of young women's sexuality, and even more strongly, a terror of fertility itself among people trying to survive in a harsh climate. Perhaps contemporary fears about scarcity, environmental pollution, and overpopulation are fueling a similar reaction against the fertility of young females that places teenage mothers among our foremost contemporary social villains.

"Girls are very dangerous at that time," wrote Maria Chono of Mexico's Papago tribe about the time of her first menstruation in the early 1860s. "If they touch a man's bow, or even look at it, the bow will not shoot anymore. If they drink out of a man's bowl, it will make him sick. If they touch a man himself, he might fall down dead." She was placed by herself in a small house, where she could not stand up, and was forbidden even to touch her own hair. After four days in this house, she was considered purified, and she was brought back to the village for fasting and ritual singing and dancing. Then when the moon was in the same phase as when it started, she was given a bath by the medicine man, a feast, and a new name. "It was over," she wrote. "I looked like half of myself. All my clothes were gone. All our dried corn and beans were eaten up. But I was grown up."

The declared dangers of young women were similar in the British Columbian and Mexican examples. But one tribe subjected the young woman to a life-threatening, four-year ordeal, while the other engaged in a month of ceremonies. The Carriers' fears brought suffering to their young, while the Papago ritualized their fears, then moved on. Contemporary America, with its many weak and inconclusive rituals, offers no clear recognition of the end of youth and the beginning of maturity. Thus, adults' fears of the young linger unresolved.

Perhaps many teenagers' recent enthusiasm for tattooing and body piercing, along with the recurrent attraction for eccentric haircuts and dress, is a reaction to our lack of meaningful ritual. Many observers of primitive societies have noted that young people look forward to rites that involve circumcision, scarification, piercing, and other painful rites because they offer visible acknowledgment of their maturity. Few contemporary parents accept and encourage these markers, as parents in tribal societies do, and most view them as signs of immaturity rather than adulthood. But they are a powerful way for young people to assert control over their own bodies and demand recognition that they are, at least, sexually mature.

Fear and hostility toward the young isn't found in every society, but it's common and some of the reasons for it are easy to understand. The simple fact is that when the young are ready to be grown up, their parents and members of their generation are not usually ready to be supplanted. The parents of young people are frequently reluctant to note the passing of their own youth, particularly in cases where the young male initiate takes on a special privilege previously held by his father.

Young people are inevitably interested in other people their own age. Such contacts allow them to measure themselves and learn about people they may know all their lives. Such curiosity is universal, and it is probably essential to the survival of the species. But in many societies this natural affinity is exaggerated by their elders' practice of segregating young people into groups whose members interact almost exclusively with each other.

Especially in agricultural societies where men control land, livestock, and wealth, young men are kept waiting at the threshold to marriage and full participation in the society for years at a time. Older men hold on to wealth and position and marry the younger women, while younger men are relegated to a separate, subordinate role in the society. They might be banished, or kept around and allowed to hunt, fight, and engage in promiscuous sex, even as they are reminded that they are less than full members of the society. In some cases, this in-between status continues until the younger men are so old they won't tolerate it, and so physically powerful they cannot be denied. Even then, there might be a ritual conflict between the generations, or sometimes real combat that results in injuries on both sides.

The practice of defining and segregating distinct youthful age groups was particularly evident in Africa. Sometimes these groups consisted of the young men who have to wait several years before the village decided to spend the money and the time required for proper initiation ceremonies. In other cases, the initiation ceremony began a period of group life. Among the Nuer of the southern Sudan, for example, initiation rituals happened every four years, and the boys took new names that identified them with the initiation group. The other members of the group were important to him during the rest of his life.

The Nyakusa of southern Africa segregated their young males even younger, at around the age of ten, when boys were no longer permitted to sleep in the same house as their parents. They went to live in a separate village, outside of the main village, though the gang of boys would appear en masse to be fed by the mothers of different members of the group. At the time the oldest boys became fifteen or sixteen, the village was closed and a new one started.

Different cultures make different rules for their youth groups. One East African tribe countenanced promiscuity for their young, and members of the group often resisted moving up to maturity and monogamy. A neighboring tribe tried to ban all sex before thirty, though it's difficult to imagine it was successful. What these (almost invariably male) youth groups had in common was that they lived in a world virtually all their own, tantalizingly on the verge of adulthood for years.

It's striking that when van Gennep wrote his book *Rites of Passage* in 1908, he felt it necessary to observe that those he called the semicivilized make a group distinction "one for which our society has no real counterpart—a division into generation or age groups."

Western history offers only a few examples of what might have been thought of as distinct youth groups. The ancient Athenian equivalent of the teenager was the *ephebe,* an upper-class young man who had finished two or more years of primary schooling and several years of secondary schooling and was supposed to be receiving military training. The young men, segregated from those of other ages, were notorious for their drunkenness, destructive pranks, and violence. The Athenians had also passed laws to forbid young people from beating their parents, an indication of a possibly justified fear of the young. We know from Greek art that youth was a quality that Greek culture celebrated. It's tempting to see them as premature Californians—in love with skin and muscle, fearful of the action these hard-bodied young men might take.

Yet, much of this seeming youth culture was passed down from the elders to the young. The symposium, or drinking party, generally consisted of men of very different ages where young men would learn to talk about adult topics, drink, and be initiated into sex with prostitutes and with the older men present. In this context, the *ephebe* seems to have been like the initiate of an "animal house" fraternity, acting a youthful hedonism and destructiveness that his elders expected and encouraged.

In ancient Sparta, it was also assumed that young people would be violent and destructive, but there it was channeled, somewhat in the manner of Mao's Red Guards, to maintain autocratic control. At age twelve or so, upper-class boys would be put under the tutelage of young men a few years older, who would, among other things, teach them how to steal to support themselves. Those who were caught were beaten severely; the lesson was not to get caught. At age seventeen, many youths became members of the *krupteia,* whose purpose was to physically and psychologically intimidate the slave population. Unlike young Athenians, whose years at the threshold involved the paradoxical qualities of invisibility and menace found in many tribal societies, as well as our own, the young Spartans at the threshold had a purpose. They were guarding the door.

In medieval Europe, there were *charivaris:* rowdy, often violent groups of young men who attacked what they saw as marital misconduct—such as older men marrying younger women and widows marrying too soon. They were fighting in the interest of their generation; it's difficult for poorer young men to compete for younger women with their propertied elders. If they could shame such rivals into fearing that they would be viewed as dirty old men, the younger men would have a chance.

Still, these are isolated examples. Throughout European history, and its extension into the Americas, family ties have always been more important than generational ones.

Today, however, it's impossible to hear about the age group practices of African and Australian tribal groups without thinking of our own. Like the African groups mentioned above, we name our generations. We are boomers, or Generation X, and we will be so until we die. We make assumptions about people's behavior and values depending upon when they come of age. Members of distinct age groups listen to particular styles of popular music, and even particular songs, from puberty to the grave.

To some degree, Americans begin to be segregated with their age-mates the moment they enter nursery school, but the separation becomes more pronounced as young people enter their teens. That is the point when parents feel that they have lost power to their childrens' peers. Young people are less likely to follow parents' agendas. Often, they don't come home for dinner, much less deign to make an appearance at family occasions. Perhaps they work at a store or restaurant where all the other employees and many of the customers are their own age. They buy products, go to movies, and listen to music, all directed at them and their friends. We encourage the creation of a youth culture, profit from it, buy stock in it. Then we're threatened by it.

By 1955, when S. N. Eisenstadt wrote *From Generation to Generation,* the seminal book on the sociology of age group relations, it was clear that youth groups were playing an important role in advanced Western cultures. Eisenstadt argued that age groups are not universal, but that they have been found in many different times and places where either of two conditions exist. The first of these is that the power of the family must be diminished. If young people cannot expect their family connections or birthright to secure them a position in the society, they must interact with those who will be coming into their full powers at the same time. Eisenstadt argued that nearly all modern societies minimize the degree to which young people can depend on their families to win position, though none more so than the United States.

Our culture created the teenager largely because we don't know what the future will be like. The American teenager as we know it today is the product of adult uncertainty about what the world would become. As we'll see, the decision to forgo the benefits of young people's labor in favor of a prolonged period of education and training grew from parents' fears that change was rendering their own skills and knowledge obsolete. A son who followed in his father's footsteps was on the road to nowhere.

Eisenstadt's second condition for the formation of youth groups is the unwillingness or inability of the older generation to pass on wealth and power to the young. If, for example, land is in limited supply, young people have to wait until they come into their inheritance, which could be at a relatively advanced age, before they can marry and lead truly mature lives. This is not an uncommon problem in agricultural societies. Alternatively, the elderly may truly be greedy and unwilling to come to terms with the needs of the young.

These two prerequisites—lack of family power and inability to share the wealth—are not mutually exclusive. The first is the main force that drives our culture's practice of prolonging adolescence and creating a nation of the young, but hints of a generational struggle for wealth are never wholly absent. The movement to restrict child labor, for example, was driven in part by a desire to shrink the labor force and raise wages. More recently, advocates of a subminimum wage for teenagers have argued that young people shouldn't receive wages high enough to tempt them to leave school. Grown-ups tell the young that they need more and more time and training to prepare for their maturity; the young aren't sure they believe it.

It's important to note that the psychological meaning of adolescence described in the last chapter is significantly different from the anthropological or sociological definition we've been considering in this one. The first is understood as an unchanging and unavoidable part of human development. The second describes the practice of many, though not all, societies, to set aside a period between physical and social maturity to prepare young people for full adulthood. Even if adolescence is a social invention, it feels very real to one who is going through it. But unlike something that is inherent in the species, such a cultural phenomenon can change rapidly, especially when it isn't doing its job.

This distinction between the psychological and social definitions of adolescence is the core of one of the most influential studies of youth ever made: Margaret Mead's *Coming of Age in Samoa*.

In 1926 the twenty-four-year-old anthropologist journeyed to the South Pacific to prove a point about American society. Her message was that G. S. Hall's vision of stormy, troubled, painful adolescence, which was virtually unquestioned at the time and remains influential today, is neither universal nor necessary. Mead needed only one strong counter-example to make her point. If there is even one society free of what were, and often are, considered the inescapable storms of adolescence, she could prove that the stress is not inherent in human development.

In Samoa, she found a society that believed controlling the sexuality of the young not to be important. Elders expected the young to be sexually active, she observed, and they didn't connect such sexual experimentation with marriage. That, at least, was the headline during the 1920s, a decade when American sexual mores were in transition and young women were the driving force for changing them. Innocent, sexy, and above all, naked Polynesians were already creatures of the popular imagination, and Mead offered what seemed to be a scientifically rigorous vision of paradise.

Mead's Samoa was a place where adolescence meant little because not very much was at stake. Children who became dissatisfied for any reason could move out of their household and into that of a relative, even a fairly distant one. Tending babies was done not by mothers but by young girls, who treated their charges carelessly. There were relatively few skills required of either sex before their ascendance to adult status. And although missionaries were present, and a boarding school as well, Mead depicted them not as forces of modernization, but rather as additional options for youth in a society where most crises could be sidestepped.

The static culture Mead found in Samoa couldn't be less like America's dynamic, fast-changing society. Mead was careful to tell her American readers that, trouble-free adolescence aside, they probably would not be comfortable in a society with so little conflict and drama. American culture, and Western culture in general, encourages individuals to be heroes of their own lives. The Samoans did not view life in terms of challenge and triumph. There was little reason to forge a strong individual identity or to struggle with one's parents. People identified with groups of people with whom they had multiple, but weak, links. The reason that adolescence was relatively trouble-free, then, was that, unlike American adolescence, it didn't really need to accomplish very much. Mead recognized that Americans needed an adolescent period to prepare their young people for life in a complex society; she simply didn't think that telling young people that they were troubled and incompetent was either helpful or true.

Since the 1980s, Mead's methods and findings in Samoa have come under attack. Mead's critics argue that the islands were nowhere near as free of conflict and strong feelings during the 1920s as her book suggests. There is little doubt that Mead had an agenda; she made little effort to hide it.

For me, the most telling and interesting criticism of Mead's work in Samoa is that she assumed that behavior that was widely tolerated was approved. In other words, she identified what was really happening—an enormous amount of guilt-free premarital sex enjoyed equally by males and females—as the expression of Samoan society's values. Mead pointed out that the young people made a pretense of hiding their sexual activities, but she did not emphasize that, if asked, most Samoans would voice disapproval.

It would be easy for an observer to make a similar mistake about 1990s North America. Statistical studies indicate that a very large percentage of teens are sexually active. This high incidence would seem to suggest that teenage sexuality is generally tolerated, though, if asked for an opinion of the phenomenon, an overwhelming majority of adults would voice their disapproval of it.

Still, the importance of Margaret Mead's work doesn't lie in the details of her ethnographic observations. Rather, it springs from her ability to imagine how cultures respond to changing circumstances and to persuade people to think about how they might change. She showed culture to be profound, persistent, and real, at the same time that she revealed it to be largely an act of human imagination. A shift in the patterns of people's expectations can bring changes more profound than can be achieved by any purely political revolution.

One needn't go all the way to Samoa to realize that cultures change and that adolescence and teenagers aren't inevitable, and may not even be necessary. We can look instead at our own past, in which young people have played a stunning variety of roles—some of which anticipated the contemporary teenage experience, many of which did not.

FOUR
Family Values

It was for your sakes especially, that your Fathers ventured
their lives upon the rude waves of the vast Ocean.

—INCREASE MATHER (1679)

More than half a century after the founding of Massachusetts, Mather,
the famous Puritan preacher, addressed a congregation of second- and
third-generation settlers about their predecessors' intentions. His message
was one repeated by generation after generation of colonists, immigrants,
pioneers, and even suburbanites: We did it for the kids.

Some parents sacrificed to assure their children's safety, others to
create a family fortune. Some were being hypocritical and did not worry
about future generations at all, though they found it useful to pretend
they did.

Nevertheless, this familiar story of what parents did for their children
leaves out something very important: what young people did for their
families, themselves, and their country. American history is a multigenerational story, in which young people have played an essential part. The
precise nature of their contribution to the culture has changed many
times over the centuries. Still, they have always played a role, and it has
always been important.

The labor of teenagers—and of preteenagers as well—has played a
very large role in the development of North America. Much of the time,

this has meant helping out on the family farm, a form of child labor that continues to be not only respectable but also exempt from most legal regulations. It has also meant working as a servant in another family's house, teaching school, logging, working in factories, apprenticing to skilled artisans, carrying cash and receipts in department stores, shining shoes, and selling newspapers on the streets.

For most of our history, child labor was not a social horror but simply a fact of life. In colonial times, it was scarcely distinguishable from helping around the house. The exploitation and abuse we now attach to the term did not exist until work moved from the home and into the factories that began to appear at the turn of the nineteenth century. Even once that happened, Americans were very, very slow to condemn it. Most developed serious moral scruples about child labor only after industrialists concluded it was inefficient, and labor unions sought to keep wages up by preventing its reappearance. For most of our history, the labor of young people—especially teenagers—has been not an abomination but a necessity. Americans had a lot of children, in part because there was so much work to do.

Many of the first English settlers of North America wanted to be able to raise their children in a setting free of the corruption they saw all about them in Europe. The Pilgrims, fleeing from England to Holland to escape religious persecution, found that the same atmosphere of toleration that allowed them to stay in Holland tempted youth, their leader William Bradford wrote, "into extravagante and dangerous courses, getting the raines off their neck and departing from their parents." His company had a difficult time surviving its first year in Plymouth. There is, however, one advantage to settling in an icy, rock-bound, savage-infested wilderness: It keeps the kids off the streets.

To mid-seventeenth-century English religious dissenters, North America appeared uniquely out of harm's way, morally speaking at least. They wanted to protect their children, but they needed them as well. The relationship between parents and children, elders and youth, is never entirely moral or emotional. It is also economic. Children depend on their parents, but in most times and places, parents have also depended on their children to help with their daily survival. This was especially true of colonial America. It was a place that needed young people, that demanded that they shoulder great responsibility, and that, on occasion, even rewarded them.

Youths in colonial American society lived in an elaborate economy of obligation. What they did had great value. It often kept the family going. Although they had the satisfaction of doing real work and making a difference—experiences that modern teenagers often miss—they were not able to achieve the autonomy that we expect such responsibility to bring. They were seen not as individuals but as members of families, and they weren't taken seriously until they were in families of their own.

By contrast, contemporary American parents delegate many of their obligations to such institutions as high school, and they abdicate much of their influence to their children's peers and the culture at large. Meanwhile, young people's obligations to their parents are weak and frequently unenforced. Yet, there is one way in which today's teenagers are closer to their parents: At least until they go to college, they live under the same roof with them. In colonial times, people in their teens had close ties to their parents, but often, they didn't live at home.

In the northern colonies, as in England, the family was the primary economic unit. It was the location of production. It was the chief policing authority. It was the only social safety net. The southern colonies were settled mostly by individuals, not families. Even there, social arrangements rested on paternal authority and filial duty.

Almost from the time they could walk, children helped. They owed their parents their labor—or the payments they received for their work—until they reached the age of twenty-one. In return, parents in the northern colonies had obligations—to see that their children had religious instruction, to see that they learned to read, and often to provide them with the training for a career. Most often, this meant apprenticeship—a relationship to a master who was, in effect, a second father. This involved a whole new set of mutual obligations, and once again, very little cash. Indeed, once money was able to trump duty, the economic system based on family values was on its way out.

It may have been enough for the first religious settlers that North America was not England or Holland. Ultimately, though, what the place was not was less important than what it was and what it might become. From the beginning, everyone saw that what it might become was rich.

The Puritans of New England had a holy vision, as did the Quakers of Pennsylvania and others who came to find a place where they could practice their religions freely. Still, even the religiously based settlements were real estate developments intended to produce a profit, even if they rarely did so. Virginia and the other southern colonies had no such overlay of piety. The settlers had come not for the kids—many had left wives and children behind—but for the money, or at least the hope of getting some.

John Winthrop, the first governor of the Massachusetts Bay Colony, anticipated the ways in which the land would change the young. While still in England, before his company departed, he drew up a list of six reasons for moving to the New World. The fifth specifically concerned the young, but it was the familiar defensive argument. "The fountains of learning and religion are . . . corrupted," he complained, and "most children, even the best wits and of fairest hopes are perverted. . . ." The third reason, while more abstract, showed greater insight into what the young would find in America: "This land [England] grows weary of her inhabitants, so as man, which is the most precious of all creatures, is here more vile and base than the earth they tread upon; so as children, neighbors, and friends (especially if they be poor) are rated the greatest burdens which, if things were right, would be the chiefest earthly blessings."

He was saying that Europe had a labor surplus while North America—if viewed in terms of European development—had a labor shortage so great it might never be filled. Indeed, it would be centuries before it was.

The promise of natural riches available to those who set out to find and exploit them is the American dream. The Spaniards began pursuing it more than a century before. Englishmen were doing it in the Caribbean and Virginia, the French in Quebec and Louisiana. Winthrop suggested that the increased value of labor would make a difference in social and political relationships as well.

He was right, though it turned out that not all such differences were for the better. Indeed, by the time Winthrop wrote, the slave trade between Africa and the Caribbean and North America was already under way. This was one response to the great American labor shortage: a massive devaluation of human moral worth. Though the Puritans did not object to slaves (so long as they were captured in a just war), this cannot be the outcome that Winthrop had in mind.

During the 1620s and 1630s, men and women known as "spirits" prowled the streets of London and other English cities. This was a new occupation, spurred by colonialism, a sort of cross between real estate promoter, recruiter, and kidnapper. They were on the lookout for young people, mostly boys and girls in their early teens, to go to work in Virginia or the Caribbean. Some of them made grandiose, never-to-be-fulfilled promises of opportunity and wealth, though more provided drink and a small bounty. Many of those the spirits recruited were orphans, homeless or without any prospects. Any change would have been attractive. The key tactic was speed and the need for a quick decision. The ship to the New World was always leaving right away. In a few instances, parents came after these people and assaulted them or filed court complaints. In most cases, the youths were already at sea, under legal bond for their labor. They had been spirited away.

Although the spirits acquired an unsavory reputation, their activities were actually a continuation of a practice begun by the government of the City of London. In 1619 it agreed to let representatives of Virginia pick 100 children from its "superfluous multitude" to go to the colony as apprentices. When some of those chosen did not want to go, the city passed a law forcing them to do so.

Sending poor people off to North America was considered to be a good both for the mother country, where they were a burden, and for Virginia, which needed all the help it could get. No less a moralist than John Donne, the great poet and preacher, celebrated this function of the colonies. "[I]t is already," he said of Virginia in a 1621 sermon, "not only a spleen, to drain the ill humours of the body, but a liver, to breed good blood."

Each indenture was slightly different, depending on the age and situation of the young person involved, but they usually required at least four years' service in return for the passage. People bought and sold the contracts, and with them, the services of the young person. The younger the servant, the longer the period of service, the calculation being that younger people were likely to be less productive than older ones and might create greater costs. As many as half of all English immigrants to the Chesapeake region during the mid-seventeenth century came as indentured servants.

It's impossible to say whether the young people were better off for having encountered the spirits. Their contracts bound them until they were in their early twenties and later. In the meantime, they had food and clothing, which they might not have had in England. Once your indenture ended, there were many economic opportunities, and a high degree of sexual freedom and individual choice in making your way in life.

Being in short supply, young women were particularly in demand, and in a position to insist on marriage. Some women married as young as fourteen, well before their indentures ended. One could always buy out an indenture, and successful men determined to marry were able to do so.

A severe shortage of workers gave servants some leverage. In one Massachusetts case, a runaway servant confessed to authorities that he had fled because he feared being punished for sexually molesting his master's ten-year-old daughter. Nevertheless, his master wanted him back; he needed the help.

Despite this extreme example, it's best to look at this system of masters and servants from the perspective of seventeenth-century family values. The family was the model for society and its only safety net. Paternal authority and filial obedience were the basis of the polity, even by those who overthrew the king and beheaded him in 1642. If you became estranged from your family, prison was the only social institution you would encounter. Those who were not part of families were dangerous because they were outside any control.

Although the family was the basic economic, political, and social unit, it was itself unstable. Disease and high mortality in childbirth, and high fertility meant an individual would have many complicated and weak relationships with stepparents and half brothers and sisters. It was similar in some ways to the contemporary "blended" families created by multiple divorces and remarriages. Malaria was common in the Chesapeake region. In the mid-seventeenth century, about half of all marriages there ended within seven years because of the death of one of the partners.

People in their teens were often bound out as servants to help support their families. Luckier ones, usually boys, were able to get apprenticeships and live in the household of a master whose authority derived from that due a father. Apprentices were sometimes reputed to be an unruly crew, whose loyalty to one another rather than their masters posed a threat to the social order.

Especially at the outset, Virginia and the other southern colonies were settled by individuals rather than by families. Indentured servitude provided a structure of authority that, in part, replaced the family. It's difficult to imagine what many of the "superfluous multitudes" recruited on the city streets would have done had they simply been dumped into the wilderness. Paying their heavy debt for the transatlantic crossing at least kept them under some control as they passed through a dangerous age.

While the settlement of New England happened very differently, the institutions of servitude, apprenticeship, and what they called "family government" kept young people dependent.

There were many indentured servants there too, but New England was settled primarily by families. Although both the church and civil government were stronger than in the other colonies, the family was the fundamental unit. Everything depended on your position within a family. The Puritans even believed that salvation ran in families and that the children of the holy had an advantage in being chosen by God, whether they showed any outward inclination toward piety or not.

Work, for the settlers of New England, was the fate of all humanity, one of the prices exacted by God for Adam and Eve's transgression. It was no less fundamental an activity than eating or sleeping. Because it generally happened in or near the household, it was part of children's experience from infancy. Parents recognized that the labor of very small children is not very productive. Still, they believed that it was good for children to become used to working. Besides, it kept them out of mischief.

Work was, moreover, not merely a punishment. One's work was part of a larger purpose: It helped sustain the family, the larger community, and many believed, God's plan.

During the seventeenth and eighteenth centuries, children would begin spinning thread when they were as young as six or seven years old and caring for animals, gardening, spinning, candle-making, and preparing food. A boy big enough to do man's work would be doing it. Those in their teens were not considered adults in most respects; they generally weren't allowed to keep any of the money they earned, which belonged to their families or paid off their indentures. Their lesser experience and strength entitled them to smaller wages. But they were not protected from work. They were too important to the economy to be idle.

Today we have an expansive idea of childhood—extending it at least to sixteen and perhaps older. The first settlers' idea of the shape of people's lives was very different from ours, though like ours, it was not entirely consistent.

They saw childhood as the period that followed infancy and preceded the development of a basic physical and intellectual competence and moral awareness. For some, childhood ended at six or seven years old, for others as late as ten or twelve. In Massachusetts, for example, military training for boys began at ten. The teenage years were part of a very lengthy, vaguely defined period of youth, which continued, practically speaking, until marriage some time in one's early to mid-twenties. Even people who held responsible political or military positions tended to be considered youths until they married.

People recognized that this period of youth brought a series of radical changes in size, sexual maturity, and intellectual capability. These changes were not occasions for crisis, but rather milestones of progress toward being a fully mature and capable person. Men at that time often didn't reach their full height until twenty-five.

It may seem ridiculous to us that a ten-year-old and a twenty-year-old could be part of the same age group. In the seventeenth century, however, both might work side by side doing the same task, and they would study in the same classroom, reading the same books with the same teacher. Physical differences between them were significant, insofar as they helped determine what jobs they would do. A big, healthy twelve-year-old could do more work than an older, less developed youth. More important, though, is the fact that the status of the youth within the society and the family did not really change during this lengthy period. The youth was deemed productive, but dependent.

Family orientation underlay this understanding of life's stages. A child was wholly dependent on the family, the youth was one who contributed, and an adult was fully capable of forming a new family. There was, officially at least, no such thing as a single person. Each of the New England colonies had laws that forbade living outside family government, and these were at least intermittently enforced. If you did not have a family, the law compelled you to establish living arrangements in a family.

The early settlers' way of dividing the life span was not as different from our own as it first seems. We make finer distinctions based on chronological age, but the lengthy period of youth persists in modern society. Children of eleven, ten, or even younger are increasingly defined as aspiring teenagers or early adolescents. They gain this less childish status, however, not because they are able to contribute to the economies as producers—as earlier generations did—but because they are capable consumers. Beyond twenty, young people become teenagers emeritus, with greater latitude than their younger brothers and sisters but with something less than full adult status. The job of such people is to go to college and farther to prepare themselves for life. This "new" stage of life most often ends at marriage. The psychologist Kenneth Keniston first took note of this emerging stage of the life span more than a quarter century ago. He called it "youth." Contemporary youth, unlike that of the seventeenth century, is seen, fairly or not, as economically unproductive. But it is still dependent.

John Cotton's 1646 catechism, from which most New England children learned, firmly established the family as the basis of all authority. "What is the Fifth Commandment?" it asked.

> *ANSWER: Honor thy father and thy mother, that the days may be long in the land which the Lord thy God has given thee.*
> *QUESTION: Who are here meant by father and mother?*
> *ANSWER: All our superiors, whether in family, school, church, and commonwealth.*
> *QUESTION: What is the honor due to them?*
> *ANSWER: Reverence, obedience, (and when I am able) recompense.*

Not only did the God-given authority of father and mother migrate upward to church and state, but the child had a debt to his or her parents that had eventually to be paid back. In practical terms, this meant that the young person's labor—whether performed directly for the family or for an outsider paying cash wages—belonged to the family. Turning over one's earnings was not optional; the young person had no right to them. The family could, if necessary, legally enforce this obligation.

The Puritans tend to be overrepresented in histories, including this one, in large part because they were so relentlessly verbal and left so many records. Their religion rested on a set of clearly stated covenants between God and man. It should not be surprising that their thinking about all sorts of relationships—especially those between parents and children—was couched in the language of contract law. Puritans sometimes wrote with great tenderness about their love for their spouses, sons, and daughters. Still, they were more comfortable enumerating the obligations of parents to their children and the debts children owe in return. In a cosmos where the devil is a very powerful force, human emotions can always be given a satanic spin that leads both parent and child astray. It was better to stick to the letter.

This fear of being misled by love was, the historian Edmund S. Morgan has speculated, one reason why these family-oriented people did something you would not expect: They sent their teenage children out of the house.

A boy who spent his teenage years living with his family was an exception in early New England, and girls were also bound out. Both boys and girls would work as servants and laborers on farms. They would live in the houses of ministers or other teachers to further their education. Many boys would spend about seven years, beginning at the age of fourteen, serving as apprentices to master artisans. In addition to their duty of making the youths competent in a trade, the masters served as stand-in parents.

Making money wasn't necessarily the motive for such transactions. Prosperous families were more likely to board out their children than poorer ones. Sometimes a family would pay to have their own children board with a teacher or collect the money their children earned as servants, and at the same time pay someone else's child as a servant in their own home.

The Puritans viewed the teenage years as a tremendously important time in a person's life, a time to make permanent choices. They seemed to believe that the discipline their child would receive in somebody else's home would be preferable to the softer, and perhaps for that reason more dangerous, upbringing they would give their own flesh and blood.

This practice, which was common in England at the time, even among non-Puritans, seems to affirm a certain psychological truth. It's easier to watch other people's children turn into adults than it is your own.

When early New Englanders spoke of a person's calling, it had a far weightier meaning than we give it now. One's occupational calling—the role you would play in the worldly community—was parallel to the second and greater calling of becoming one of God's elect. This second calling was believed during the seventeenth century to happen quite late, usually in one's twenties.

Because the Puritans believed nearly all girls would have exactly the same worldly calling as wives and mothers, they weren't subject to the proto-identity crisis their brothers often had to undergo. Likewise, for the majority of young men who spent their lives farming, their searches for a calling were less anxious those of young men in town. It was not proper for a parent simply to dictate what his fourteen-year-old would become. The Puritans believed that each person had unique, God-given talents. The teenager and his parents would pray and look for interests and abilities that suggested the path of their true calling. It was a tense and anxious time, similar to today's nerve-wracking college admissions ordeal. False starts were tolerated, sometimes more than once, when it was clear that the boy had not found a proper calling.

This was the case with young Sam Sewell, a member of one of the founding families of Massachusetts, who was in 1694 apprenticed to a shopkeeper. By the next year, he was back at home, causing his father to pray constantly "for Samuel to be disposed to such a Master and Calling, as wherein he may abide with God." He then became apprentice to a Captain Checkly, another merchant. There, young Sam had problems with his work because of "the numerousness of goods, and hard to distinguish them." He argued that books, with their clearly marked names and prices, would be easier for him to handle. Following more prayer, more sleepless nights, and a negotiation with Captain Checkly, Sam was finally placed with a bookseller.

Once parents and children navigated this difficult passage, the young man left home for another family. His master was a second father, who had the right to demand his services twenty-four hours a day, seven days a week. The master's wife became, if not quite a mother, a key figure in the apprentice's life, because she was responsible for his meals, the circumstances in which he lived, and often the clothes he wore. Often, the apprentice shared a bed with the master's own children. Although this does not seem an ideal arrangement, it was at least an indication that the master was not treating his apprentice any worse than his own children. Some apprentices had to sleep in a pile of filthy straw on the floor of the master's unheated workshop. This permitted a measure of independence, as apprenticeship went, but little comfort.

Apprenticeship was also a form of education; in some of the colonies, masters also had more bookish responsibilities. They had to teach their apprentices how to read and write, if they didn't know already. In Massachusetts, a law required masters to make sure that their charges studied their catechism. There, the apprentice should have known how to read because, since 1642, the law had required that parents teach their children how. In 1647, citing the need to counteract the influence of "that old deluder, Satan," a law compelled towns to run schools that would teach children to read well enough to understand the Scriptures.

Learning played a big role in Puritan beliefs. To be ignorant was almost certainly to be sinful. Even so, some towns were reluctant to spend the money and make the effort that this basic education required. This became particularly true after the compact villages that were the earliest settlements became dispersed, as farmers preferred to live on their acreage rather than in the village. Some towns dealt with the problem by convening school only when the General Court, which was a combined legislature and judiciary, was in session. It wasn't in session all that often.

If towns in the most education-obsessed of the colonies took such a casual attitude toward this obligation, in Virginia and other southern colonies, it was scarcely happening at all. One problem there was that there had never been villages with enough people nearby to support a school. Eight decades after the colony's founding, concerns that the colony was producing a generation of ignoramuses prompted a plan to build county seats, largely so there could be a place for a school.

For the very wealthy, especially the large landowners of the South, sending their children to England to study was an option. To get them ready for higher education, they would often hire tutors. Whether the college to which one aspired was Harvard or Oxford, the key to the preparation was knowledge of Latin, then still the language of higher learning.

Teenagers also showed up in other educational settings. They might be delayed students in reading school, where both boys and girls mastered the rudiments of that subject, or if they were boys, in writing school. They might also have been the teacher. Youths often took off time from grammar school or college to make some money teaching elementary subjects to classes of various ages.

For a contemporary teenager, most kinds of work are forbidden or highly restricted until the age of sixteen. Thereafter, work is supposed to fit around one's schooling. In the seventeenth and eighteenth centuries, work was primary and schooling fit in between the work. Schooling tended to occupy short, intense periods of people's lives when there was nothing useful to do. Boys most often went to school in midwinter, when both they and the men who taught them had little to do on the farm. Even students who went on to higher education were often late starters, learning reading and arithmetic in their teens alongside children a decade younger, at the same time they were performing a grown-up's work on the farm.

Because attendance at school was intermittent, what counted was not the age of the student but the books mastered and the skills acquired. It was possible, if a boy applied himself, to pass through the different levels relatively quickly. "Remember," wrote a Boston father in 1672 to his son entering Harvard, ". . . that tho' you have spent your time in the vanity of Childhood; sports and mirth, little minding better things, yet that now, when come to this ripeness of Admission to the College, that now God and man expects you should putt away Childish things: now is the time come wherein you are to be serious, and to learn sobriety, and wisdom in all your ways which concern God and man." The son was only fourteen years old. He was entering this serious realm along with other young men who ranged up to a decade older. They were all seen as youths.

Apprenticeship was more important in New England and the middle colonies than it was in the South, where skilled craftsman frequently arrived as indentured servants and slaves served both as artisans and un-skilled laborers. The growth of such large cities as Philadelphia, New York, and Boston encouraged the growth of the crafts and of apprentice-ship, and of institutions like apprentice night schools and libraries. It's too much to say that they were early teenagers. But they were visible and remarked upon, though often in a tone of condemnation about what's become of young folks nowadays.

As in a family, the relationships between masters and apprentices involved mutual obligation. The chief duty of the master was to train the young man in, as apprenticeship indentures often expressed it, the "art" and the "mystery" of the trade. By "art," they meant skills refined through many years of practice. "Mystery" referred to specialized knowl-edge—in effect, trade secrets.

The obligation of the apprentice was, essentially, to do whatever the master required. An apprenticeship indenture was a transferable that passed to the master's heirs at his death. Young men often found them-selves legally bound to masters who did not know the craft they promised to teach. Sometimes they had to take masters to court to force them to fulfill the conditions of the apprenticeship.

Even when the original master hadn't died, he might still be unavail-able, as in the cases of the four Philadelphia apprentices who complained that they were not being taught their trade because their "Master has been for some time confined in Gaol, and when at Liberty is very Seldom at Home as can be proved by Numbers of his Creditors."

There were also many cases in which the master simply decided to change the work he did, leaving his apprentices high and dry. The Puritans may have believed that each person had a fixed calling, but even in New England, and more so in New York and Pennsylvania, a distinctly American sort of occupational restlessness took hold early.

The work apprentices performed often had nothing to do with the trade for which the young man was preparing. Indeed, there was an expectation that the apprentice would serve as a common drudge for some years before he would even begin to study the trade. Some masters seemed to have abused this expectation and never taught the craft. In their defense, masters said they feared that the apprentices would run away and find work elsewhere, and they needed to get some value from the apprentices before they did.

This conflict shows that even though the colonial apprenticeship system followed the English model, conditions here made the practice very different. Often, the master himself did not have anywhere near seven years' worth of arts or mysteries to impart. The American colonies were essentially without guilds to organize and police the system. Although the works of a handful of fine colonial cabinetmakers and silversmiths were the first masterpieces of American art, colonial craftsmanship was, at best, uneven. Almost anyone could declare himself to be a master and take on apprentices, provided that their fathers were willing. Indeed, men would sometimes change occupations and then take on apprentices, even before they had learned the trade themselves.

Apprentices were not supposed to leave their masters. Ships weren't supposed to let them board, taverns weren't supposed to serve them. And until the early eighteenth century, many places had laws requiring that runaway apprentices be publicly whipped. Later, punishment of a runaway apprentice was the responsibility of the masters, who in many cases offered rewards for the capture and return of the apprentice. Indeed, runaway apprentice advertisements were a mainstay and revenue producer for the colonial press. They are valuable and interesting for their detailed description of the apprentice's appearance, habits, demeanor, character, and of every article of clothing the apprentice possessed.

Especially later in the eighteenth century, the clothing described became increasingly elaborate, including colorful jackets and patterned waistcoats. For the first time, young people were dressing differently from most adults, and their bright, luxurious clothing seemed to some an act either of defiance or degeneration. Elaborate apprentice dress provoked warnings—especially from Quakers and Puritans—that youth was becoming decadent and effeminate. Americans seem always to view the spectacle of young people spending their own money as a symptom of social decline.

While the peacock propensities of apprentices may have been exaggerated by those deploring them, the trend to better clothing suggests a power shift. Apprentices were able to win more concessions from their masters, in the form of clothing or cash, because the masters were prospering and they needed the apprentices. One indication is that even though the advertisements' descriptions of the apprentices themselves were often insulting, the masters were willing to risk money on the increasingly unlikely possibility that the apprentice would come back.

In this, as in so many other cases, Benjamin Franklin is an exemplary American. He was apprenticed as a printer to his much older brother James. Being apprenticed to a relative was quite common—apprenticeship was only a logical extension of the family. Franklin didn't like having his brother as second father and their relationship was stormy. When James was jailed because of some of the things he said in his newspaper, Franklin continued the newspaper, for which he wrote widely noticed articles and ran the printing business. Later, when his brother was out of prison and expected Benjamin to return to a subservient role, they clashed. At the age of seventeen, Franklin left Boston for New York, where he could not find work, then for Philadelphia. He won his job there by showing he could set type. The keys to success for a runaway apprentice were to run far enough—and to be able to do the job.

I have talked so far about classic artisans' apprentices. This familylike method of training was also used to prepare people for both the professions—as a doctor or lawyer—or the upper reaches of trade and commerce.

Fathers paid dearly to place their sons with successful merchants and traders in the largest cities. Typically, they made a large payment for the initial placement and would provide further sums for clothing and schooling as the master specified. Being apprenticed to a distinguished merchant was a valuable endorsement. The fathers who paid the bills expected their sons to meet influential people and build reputations that would allow them to start businesses of their own. Sometimes the young people would marry the master's daughter and inherit a successful enterprise. This method translated country wealth, derived from large landholdings yielding rent and agricultural products, into urban wealth, based on commerce and finance. Poor but deserving apprentices had little chance of being placed with the most successful and powerful businessmen.

Apprenticeship was also as a way of dealing with the poor. Often, very small children, some as young as a few months old, were bound out for apprenticeships that lasted until adulthood. Such children included orphans and those whose parents were unable to care for them, or couldn't afford to do so. Binding children out for apprenticeships prevented the children from being a public burden, and it was argued that the children would be far better off in the setting of a good family.

This was the one sort of apprenticeship that involved a large number of girls. Most of them were apprenticed into the trade of "housewifery." The families that paid for the indenture purchased the services of a household servant until she reached eighteen years of age. Poor boys were likewise apprenticed as farmers, another trade that had little value unless you had the capital to acquire land. While merchant apprentices moved prosperous country boys to the city, poor apprentices most often started out among the abandoned people of the cities and moved to the country, where unskilled labor was more valuable. Aspects of this form of apprenticeship survive today in the practice of placing abused or abandoned children in foster homes. The biggest difference was that the poor apprentices became, in effect, the property of their masters, and once they did, the authorities seldom checked on the young person's situation.

Toward the end of the century, yet another variation on this system arose, what might be termed false apprenticeship. This was a situation in which a master used apprentices as cheap labor in a workshop that was evolving into a factory. Thus, a clockmaker would not impart all the art and mystery of that craft, but would rather teach the apprentice how to make a single subassembly or to craft a part. Of all the people in the shop, only the master knew, or would ever learn, how to make a clock in its entirety. Soon there would be factories in which nobody could make a finished product from the raw materials.

This was clearly an efficient way of making clocks and many other products. It happened gradually and quite subtly, and it sometimes took years for the youths to discover that their apprenticeship would never teach them how to be independent artisans. Instead, they had become part of the first generation of industrial workers. There were things to learn in this new setting. The apprentices might observe how a factory works, and the increasing mechanization of production produced new opportunities for mechanics. Still, the essential bargain of apprenticeship, based on a paternal responsibility of the master and his concern for the youth's welfare, was abrogated.

Apprenticeship works best in a relatively static economy, marked by strong social ties and slow technological change. What is surprising about apprenticeship in America is not that it ultimately died, but that it took more than two centuries to do so.

One reason may be that it was psychologically sound. Apprenticeship at its best provided youths with clear goals, with skills they could master to achieve tangible results. It provided them with role models, a realistic sense of what their lives might become. It established relationships that sometimes endured and helped shape the life of the young person ever after.

Almost from the time apprenticeship withered as a principal method of education and training, there have been calls for its revival. Today is no exception, and it's easy to see what appeals to its advocates. It would counteract the bureaucratic environment of the high school by establishing personal relationships, in which masters and apprentices have clear responsibilities to one another. The influence of the teenage peer group would decrease, while cross-generational ties would strengthen. The apprentices would learn discipline and good work habits. The work that the young do would be hands-on, not abstract. They would be able to see the impact of what they do. And young people who lack a social network to let them know about the availability of jobs would establish relationships that would help them throughout their careers.

Recent proposals have called for this apprenticeship to be a supplement to high school so that students would graduate with both a high school diploma and a certificate of competence in a particular field. Nearly half of Germany's secondary students participate in such a program, though that country's social contract of high wages, high unemployment, high taxes, and high social subsidies is clearly different from ours.

Whether such a hybrid program can work in an American context isn't clear. Much of the appeal of apprenticeship is nostalgic. It assumes a less dynamic world, one in which skills are passed from hand to hand. Today's youth cannot afford to respect the skills of their elders too strongly; they will be obsolete very soon. The idea of apprenticeship also assumes community-minded business, happy to subsidize the training of young Americans, even though workers in Asia or the Caribbean cost one tenth as much. The apprenticeship system was already under siege two centuries ago because there were too many opportunities and too much change. There still are.

Sociologists make a distinction between societies that are particularistic—in which individuals are assigned a specific role, usually on the basis of family and birth order—and those that are universalist—where all roles are, at least theoretically, open to all. North American society is classically universalist, where money talks and everyone takes standardized tests.

Still, within our own families, each of us plays roles whose origins we do not remember. We incur obligations that we cannot pay off in cash, but only through many years of devotion and care. We tend to view such relationships in terms of love, not of economics, though many families have recently discovered the expense of replacing a mother's care.

An economy based on obligation is workable in a closed society with fixed institutions. Americans realized during the eighteenth century that their situation was open-ended. A person could run away from obligations. Cash is portable and recognized by all. Fathers increasingly understood that the greatest favor they could do for their sons was to release them and let them make their own way.

In his classic study of the first four generations of Andover, Massachusetts, Philip Greven described a process repeated over and over again in towns of the North. The first generation, present at the first allocation of the town's land, had more land and more work than it could handle. Landowners left large chunks of their land undeveloped because they were unable to work it. But the farmers' wives had children about every other year for a period of twenty years, while both deaths in childbirth and infant mortality declined. Most fathers had enough land to pass on to all their sons, and often their daughters as well. While the fathers often made specific promises of land, which brides' fathers demanded before they would let a marriage proceed, they did not cede control until fathers, and often sons, were quite old. The members of this second generation seemed satisfied to wait.

By the third generation, most fathers did not have enough to offer their sons to keep them nearby and dependent through much of their adulthood. There were opportunities to get in on the beginnings of other towns and new careers emerging for businessmen and merchants in the cities. More often, only one or two sons would be given the land and allowed to take control of it at an earlier time of life. By the fourth generation, looking outward for your future had become common and expected.

During the seventeenth and early eighteenth centuries, this process primarily affected people in their twenties and early thirties. Youths in their teens remained subject to their parents, though often separated from them. But the changes that were coming to the society—the growth of cities, religious revivals, the beginnings of industry, the growth of education—would have enormous impact on teenagers. This was a young country. By 1776, half the population of the newly conceived United States was under sixteen.

FIVE
Declarations of Independence

Nothing hurts the affections of parents and children so much, as living too closely connected, and keeping up the distinction too long. Domineering will not do over those, who, by progress in life have become equal in rank to their parents. . . .

—THOMAS PAINE, *The Crisis* (1776)

Was Thomas Paine, the great propagandist of independence and revolution, really venturing into family psychology here? The date of this passage would suggest otherwise. As we have seen, colonial Americans saw the family as the basis of government authority. Paine, a skilled polemicist, would naturally resort to the terms people understood best to make the case for asserting American independence. Besides, the argument that the American colonies were mature enough to stand on their own had been made before, and it permeates the Declaration of Independence.

Nevertheless, such assertions of maturity, independence, and the right to make one's own way would not have had so much political impact had they not reflected what people were experiencing in their own households and their own lives. Throughout the eighteenth century, the claims of youth had become stronger and stronger. Increasingly, young people found God without needing to use their family connections, which the Puritans thought were so important for salvation. They educated themselves, often avoiding the family-based relationship of an apprentice. They explored new opportunities that were becoming available both on the frontier and in the growing cities. The lives they led looked back in many ways to medieval Europe, but in other ways they were adjusting to a modern pace of change.

They stood, as the young so often do, at a threshold. Their power and their responsibility weren't yet clear. Still, the late eighteenth century brought with it many personal declarations of independence. Sons weren't waiting to inherit property but rather looking to make their own way in a world they understood better than their elders did.

I'm not trying to argue that the American Revolution was a youth uprising. It was obviously more than that. Still, it was the climax of more than five decades in which the claims of youth were asserted and accepted in many areas of life.

One of the longest-running arguments about the American Revolution is whether it was "conservative" or "radical." Was it an effort to preserve property, business, and an autonomy to which the colonists had become accustomed? Or did it represent a fundamentally new idea about how people should govern themselves?

The willingness of the radical Thomas Paine to cast the revolution in terms of generational conflict suggests a slightly different perspective. The claims of a younger generation are essentially conservative ones— they want what their parents have. But they inevitably see things differently from the way their parents do. Thus, the young see their desires as only reasonable, while older people feel the world is turning upside down.

There was one thing that eighteenth-century Americans couldn't help noticing about their teenage children. They were big.

American-born youngsters were, on average, growing much taller than their counterparts in England and elsewhere in Europe. Moreover, they were reaching this larger size more rapidly. By the middle of the eighteenth century, American males had reached an average height of five feet, seven and three-quarters inches. Their English counterparts were just under five feet, five inches, a very noticeable difference. While the height of upper-class Europeans was comparable to that of Americans throughout the nineteenth century, European stature on average did not catch up with that of Americans until the mid-twentieth century. (Today Norway and the Netherlands lead the world at five feet, ten inches, while the United States is tied with England at just under five feet, nine inches.)

While statistics can't provide a generation by generation measure of increasing stature, they suggest why, from earliest times, American children shocked their parents with their unexpected size and power. This phenomenon is still evident among contemporary immigrants, especially Asians. One often sees frail, small-boned parents dwarfed by their high-protein progeny.

Increased stature is strongly tied to with ample nutrition and the absence of environmental stress and disease. The earlier onset of menstruation discussed earlier is also associated with these factors, though not quite so strongly. An unprecedented aspect of this great American growth spurt was that—with the crucial exception of the slaves—there was no difference in stature associated with social class, occupation, or place of residence. All were growing together. Even slaves, who were abnormally small as children because of infant malnutrition, typically experienced an adolescent growth spurt that left them at the short end of the range of normal American height.

The evidence of increased stature demonstrates that a high standard of living has been part of American identity from the beginning. My point here, though, is that young Americans quite literally embodied change. Parents could see that their children were different from themselves at their age. Even at the time, this was seen as a result of the rich resources and opportunities available in North America, but it was also frightening. In at least the material realm, "doing it for the kids" had worked. In a deeper sense, though, there was cause to worry about whether young people would be controllable. Would they accept their parents' values as their own, or would this new place change them into an altogether different sort of people?

Indeed, there was anxiety, especially in western Massachusetts, that young people would reject English civilization and take up the lives and ways of the Indians. Some young people captured by Indians continued to live as Indians and resist those who attempted to "rescue" them. Colonists nervously acknowledged that the Indians knew more than they did about the place where they were living, and they were capable of appearing from nowhere, often to threaten them. They weren't the last parents to fret that their children might turn into what they most feared.

With few exceptions, however, the children turned out as they usually do—like their parents in most essential ways, but different in a few important ones. By the end of the century, young people were not nearly as subservient to their families or as dependent on familylike institutions, such as apprenticeship, as they had been earlier. They were no longer transplants.

Among the things that changed the lives of eighteenth-century American youth were new kinds of schooling, emergent patriotism, and for some, the rise of social life in the cities. But the first and most lastingly important was religion.

Religion does not seem, at first glance, to be a likely arena for the assertion of youthful independence. An embrace of the faith of one's fathers seems a reaffirmation of paternal authority. Nevertheless, religion must appeal to youth in order to survive. The way in which young people understand and express religious principles changes them, but it also changes the religion itself. Religion must inevitably be an issue as young people exercise their newfound ability to think for themselves and examine what others have told them. Moreover, religion speaks to some of the big issues of life—human purpose, the nature of the universe, death, love—with which young people have perennially grappled. To be or not to be is, after all, an interesting question. Even today the emotional dimension of religion draws the young.

On August 13, 1993, half a million young people, most of them in their teens, converged on Denver for a once-in-a-lifetime experience. There were several large events in a packed stadium, and 110,000 marched fifteen miles for an appearance by the main attraction. An estimated 20,000 suffered from dehydration, altitude sickness, and other ailments at the event. It was bigger than Woodstock, at least as emotional, and arguably more dangerous. But the person they had come to see was not a rock star. It was Pope John Paul II, and the event was the Catholic Church's World Youth Day, the first held in North America.

Many parents of those who attended were probably relieved that they had gone to see the pontiff rather than some depressive heroin addict in a flannel shirt. Compared with a rock festival, it unquestionably featured fewer drugs, less sex, and a lot more nuns. Yet, in its intensity, its occasional chaos, and even its overall good-naturedness, it appeared to many virtually indistinguishable from other youth extravaganzas. Those who attended may have had some values and beliefs that differentiated them from their peers, but they were part of the same culture.

One need only eavesdrop on the semiprivate conversations on the Internet to see that core Christian issues such as the existence of God, the truth of the Bible, the existence of evil, and the nature of faith are debated.

Admittedly, much of this contemporary dialogue is generated by the fundamentalist religious revival among youth that is one of the most lasting, if least recognized, survivals of the 1960s. The association of this movement with creationism, antiabortion activism, and other causes of the religious right tempts us to label such strong expression of youthful religiosity a conservative phenomenon.

But we need only look back to the 1970s to see that religious feelings among the young are not always conservative. During that decade, the Unification Church, the Hare Krishna movement, and other religious cults actively recruited teenagers and alienated them from their parents. The political beliefs of the leaders of some of these organizations may have been right wing politically, but their impact on family life was deeply subversive.

It's worth keeping these contemporary and recent examples in mind when looking at young people's embrace of religion in the mid-eighteenth century. It was labeled a revival of faith, and the doctrines being preached—notably a preoccupation with the torments and the likelihood of hell—were probably more "conservative" than those held by the most zealous Christian fundamentalists today. Yet, the result was, in the long run, subversive to the status quo and very likely a step toward revolution.

What has become known as the Great Awakening began in New England as an effort to reinvigorate Puritan teachings more than a generation after they had lost much of their power within the society. Its intentions were not conservative but reactionary, an attempt to turn back the clock to a more religious time.

There was, however, a major difference between the earlier religion and its revival. While Puritanism was itself an act of rebellion against the established church, in New England it resided within a fixed, stable family structure. The father was not only the head of the family, he was its chief teacher of religion. He was exhorted to regularly go over the catechism with each of his children to determine that they both knew and understood the doctrine.

Children were welcomed into the church because of their family connections. There was little expectation that they would be ready for conversion—full membership in the church—until they were well into adulthood. The ages between twenty and forty were judged the most suitable to answer Christ's call. In the early years, men typically did not become full members of the church until they were in their mid-thirties, while women tended to be converted several years younger. There were cases of precocious callings, but the faithful viewed them suspiciously. Puritanism was a highly intellectual faith, and ministers often remarked that people are not truly capable of mature judgment and understanding until their twenties.

Moreover, when the Puritans spoke of "conversion," what they meant was converting the offspring of Puritan families into full church members. Those who were not believers had to spend time in Puritan churches, but there was little expectation that such people would fully embrace the faith.

The Great Awakening was, by contrast, something far more familiar to us today. It was evangelical, an attempt to get new people into the church door. While it drew from the difficult teachings of Calvinism, it was highly emotional. The vivid threats of hellfire and eternal torment were balanced by the promise of an ecstatic, transforming oneness with Christ. You didn't have to be born into the right family. This spiritual experience was as available to women as to men. And it might happen at any moment. Young people could now trust their emotions. Their parents need not worry that youthful conversion was a trick of Satan. People in their teens, and even younger, participated in the Great Awakening. They were among the most enthusiastic of the converted. Indeed, the struggle for the souls of youth was sometimes cited as the reason such a religious revival was necessary.

"Licentiousness for some years greatly prevailed among the youth of the town," wrote Jonathan Edwards, the leading figure of the Great Awakening, about Northampton, Massachusetts in 1730. Young people were, he said, "very much addicted to night-walking, and frequenting the tavern, and lewd practices, wherein some, by their example exceedingly corrupted others." This passage is one of the first descriptions in American history of what would become youth culture. Already, the tone of high disapproval and elder exasperation was firmly in place. "Very frequently," he wrote, "conventions of both sexes, for mirth and jollity, which they called frolics, [spent] the greater part of the night . . . without any regard to any order in the families they belonged to: and indeed family government did too much fail in the town."

The Northampton revival of 1734–35 put an end to this moral chaos. Edwards cited the conversion of one young woman, "one of the greatest company-keepers in the whole town." The impact of her sudden and complete change of path and acceptance of God was, Edwards wrote, "almost like a flash of lightning, upon the hearts of young people all over the town, and upon many others."

Family government may, as Edwards said, have been failing, but his movement did little to restore its legitimacy. By encouraging young people to see themselves in personal contact with God, he bypassed the role of the father as spiritual authority in the household. Because one's religious calling was viewed as the most important thing in life, young people who experienced religious conversions saw themselves as equal to their parents, perhaps superior. Edwards cited religious prodigies as young as four years old, an indication that he was not concerned with intellectual rigor or mature judgment. Many more of the converts were in their teens. Their motive seems not to have been to rebel against their parents, but to find an intense religious experience.

Still, righteousness has a way of curdling into self-righteousness. Those who were skeptical of the Great Awakening focused on its impact on the relations between young people and their parents.

Charles Chauncy, a Boston minister who was worried about "the bitter and Rash judging," familial tension, and lack of charity that the movement has brought, quoted, in a 1743 work, a letter he said he had received from a friend: "There was a young Woman about 15 Years of Age, who fell under this Conviction, and for about four Hours together she, in this manner, exhorted. At first she began with her Father, and told him, she could see the Image of the Devil then in his Face, and that he was going Post-haste down to Hell; and that all the prayers he had ever made in his family were nothing but Abomination in the Ears of the Almighty, and that all the Counsels he had even given her had no better a Tendency than to instruct her how she should please the Devil; and that both he and his wife were no better than the Devil."

We might speculate that father-daughter relations within this household had probably deteriorated some time before the young woman found religion. What is important is that embrace of religion gave this fifteen-year-old a principled basis for rejecting her parents—or at least for throwing a pretty dramatic tantrum.

It's worth mentioning, too, that religion, especially when it promises direct contact with God, has often been a powerful vehicle of expression for women. Christianity may be patriarchal, but it declares that both sexes are equal in heaven, if not on earth. A religious movement like the Great Awakening, which was critical of things as they were, offered an opportunity for women to exercise a measure of independence and of leadership. Young men had plenty of opportunities to get out of the house and test their competence. The church was almost the only public institution to give females an important role, if not full equality. It's not surprising that women in their teens embraced this respectable way to assert their piety and autonomy at once.

It is not so clear whether the Great Awakening was the cause of youthful expression of autonomy or simply a vehicle for it. Indeed, Edwards's description of the wayward youth of Northampton suggests that some sort of rebellion was under way before religion emerged to absorb and amplify its energies. Was what Edwards saw a menace or a market?

The answer seems to be both. Young people had been showing signs of independence, which would become more pronounced as the century wore on. As parents had less control over the economic status of their teenage children, they had less control over their conduct. Records suggest that out-of-wedlock pregnancies increased to a level not seen again until the twentieth century.

Bundling, going to bed clothed with a member of the opposite sex, was a popular evening's amusement. This was a more or less respectable activity, though if the woman became pregnant, the man would be excommunicated from the church. "She'll sometimes say when she lies down" went a popular ballad about a bundling maid. "She can't be cumber'd with a gown, / And that the weather is so warm, / To take it off can do no harm. . . ." Advice-book authors were still instructing young people to always bow before their fathers and mothers, but parents themselves were learning to be tolerant of their children's behavior.

The emergence of churches that embraced young people and took their experiences and their emotions seriously was probably inevitable. While the Great Awakening was local and short-lived, the transformation of churches by youth and youth by churches would continue for more than a century. Toward the end of the century, it was almost expected that a young person would experience some sort of a religious crisis, much as Erikson has conditioned us to expect an identity crisis.

Young people used these crises in many ways. Devereaux Jarratt, who was born in Virginia in 1732, was orphaned as a boy and bound out as a servant. "I was assaulted with very uncommon trials," he later wrote. "Sorrow, trouble, and perplexity, continued long and painful— perhaps for twelve months." This led, he said, to his conversion, which placed him on a path to divinity school and what became a successful preaching career.

More than fifty years later, Joel Winch, a young man in Vermont, had a messier and somewhat less literate conversion experience. As a youth, he wrote, he thought "much about dying" and would often "git alone and cry." The Lord "broke into" Joel's soul at fifteen, but "the Congregationalist professors would take no notis of that which the lord had don for me." Older people in his congregation believed that teenage conversions were unreliable. After his apprenticeship was over, Joel backslid. He tried with his friends to start a witchcraft scare by opening people's doors, making noises, putting cows in pigpens and pigs in cow pens, and throwing carts down wells.

Eventually, Joel's religious rebellion became a public issue. He became a Methodist, but was ordered by his former church to "come out from among" the Methodists "and be separate and tutch not the unclean thing." Joel determined "to stand fast in the liberty whare in Christ had made me free." He continued to attend meetings of his old church, asking embarrassing questions. The pastor of the church sought to excommunicate him, but the congregation eventually came down on Joel's side by allowing church members to join other churches.

Joel Winch appears to have been a confused and deeply irritating young man. Still, in his erratic way, he seems to have realized one abstract liberty recently enshrined in the Bill of Rights—freedom of religious expression.

Joseph Kett, the influential historian of youth in America, has seen in the latter half of the eighteenth century the first appearance of aspects of the phenomenon we now call adolescence. He argues that earlier young people experienced a long period of semidependency, in which they made relatively few important decisions about their lives. By the mid-eighteenth century, however, new economic opportunities were enticing young men away from their families, young people were making religious commitments, and they exercised more control over their social life and marriage than before. The result was that more decisions had to be made in a shorter time. Moreover, some of these decisions—to leave the farm and strike out for a new part of the country, or to become a minister rather than to follow your father's trade—meant rejecting parental values.

What makes this issue complex is something we often find in society's dealings with young people. While parents' declared values might remain conservative, they are often willing to allow, or even encourage, bold action by their children. Though a father had a right to all his son's earnings until the age of twenty-one, many allowed the son to amass enough capital in his teens so that he could "reject" his father and go off on his own.

The role model for many young people at the end of the eighteenth and beginning of the nineteenth century was Benjamin Franklin. While today we remember him for the rather stodgy advice he peddled when young—the penny-saved, stitch-in-time aphorisms of *Poor Richard's Almanac*. In his own time, though, he was admired more for the life he had lived. He was a runaway apprentice, an entrepreneur, an innovator of social and civic institutions. Alone among the nation's founders, he was an exponent of city living, and he celebrated what most of his fellow founders feared: an urban future for Americans.

Settlement of the western frontier—meaning at this stage the Appalachian region and the areas beyond to the Mississippi—was the most obvious focus for new opportunity. Franklin was the pioneer of another frontier created by urbanization. At the time of the revolution, Philadelphia was already second only to London in population among British cities, New York was growing rapidly, and Boston, Baltimore, and Charleston were centers of commerce and culture. By the end of the century, a young man leaving a small town in New York, Pennsylvania, or New England was as likely to be seeking his fortune in a big-city countinghouse as on the frontier.

Anyone who visits Philadelphia quickly discovers that Benjamin Franklin seems to have founded nearly everything. Libraries, postal service, a hospital, fire protection, insurance, education, and street paving all seem essential to city life, and Franklin involved himself in all of them. To call him a founder, though, is a bit misleading. He was more of an organizer and propagandist for voluntary organizations that provided these services. In most cases, his fellow founders were leading people of the community—though the fire brigades quickly became dominated by young people and a few were indistinguishable from youth gangs.

In addition to providing institutions that increased the appeal of urban life, the civic activism that Franklin represents served as a model for voluntary organizations of all sorts. Increasingly, young people were taking things into their own hands. At colleges, for example, they began to found extracurricular organizations that supplemented, and in a few cases threatened, the offerings of the university itself. Apprentices' libraries were ultimately forced by their constituents to offer books, especially novels, that their young clientele actually wanted to read.

Franklin's autobiography, which was one of America's bestselling books for more than three decades, was a chronicle of chances taken, opportunities seized, and new ideas realized. He was a living example that old patterns and paternal expectations no longer applied. He proved it possible to become a sort of person nobody had expected.

Franklin further contributed to the increasing independence of American youth by proposing and advocating a novel educational institution: the academy. Although hardly any that were founded embodied the fusion of serious learning and practical results Franklin envisioned, they made more secondary schooling available to more people than ever before.

The struggle between academic and practical schooling is the longest-running issue in American education. During the first century of American settlement, there was little resistance to the antique Latin grammar school or to the liberal arts education to which it led. William Penn had suggested that study of nature and other scientific pursuits could teach students the same mental discipline as linguistic knowledge, and have some useful outcomes as well. He did not, however, try to put these ideas into action in the colony of which he was proprietor. Perhaps because grammar schools and colleges served relatively few, and apprenticeship served many, there was little movement for innovation or reform.

Still, there were signs of a particularly American form of youthfulness emerging. Young men from the colonies who went to England for schooling appeared restless and easily bored. Although the first reaction was that this represented a failure of the young people, others took note of their "volatile genius" and called for schooling suited to the American temperament.

"Something seems wanting in America to incite and stimulate Youth to Study," Franklin observed. Americans already saw themselves as pressed for time. "As to their STUDIES," Franklin wrote, "it would be well if they could be taught *every thing* that is useful and *every thing* that is ornamental: But Art is long and their Time is short. It is therefore proposed that they learn those Things that are likely to be *most useful* and *most ornamental*. . . ."

His solution was a privately operated, though publicly endowed, institution that would produce a graduate both competent and cultivated. Franklin yoked mathematics to accounting, history of commerce and map reading to the traditional ancient history, oratory to morality and leadership, mechanics to gardening and grafting. And he would teach grammar by reading some of the best contemporary writers of English.

Franklin's proposals don't seem revolutionary to us now. Such a smorgasbord of the traditional, the practical, and the novel prevails in American schools at the end of the twentieth century. He proposed it in 1750, and it was immediately influential, although it took two centuries for it win the near universal acceptance it has now.

Franklin's was a vision of democratic education, a school that fits all without prematurely foreclosing anyone's life choices. It could prepare youths for a college education, but it could also serve as an alternative to college. It is a vision that continues to distinguish American schooling from that of France, Japan, Great Britain, and countless other places. Frankly a hybrid form, Franklin's academy had motley progeny, that reflected very different ideas about what schooling ought to accomplish.

The academy that Franklin founded in Philadelphia evolved into the University of Pennsylvania, a school whose nature converged with Harvard, Yale, and the other colleges to which it was intended to be an alternative. The first academy founded on Franklin's principles in New England was Philips Andover, which evolved into a college-preparatory school only subtly different from the grammar schools it sought to supplant.

Neither of these distinguished survivors of Franklin's vision offers much insight into the academies young people of the eighteenth and nineteenth centuries experienced. Despite its claims to practicality, Franklin's academy was almost as much an ideal as Plato's, and few of the real academies ever came close. Very few were of high quality. They were a continuing improvisation; most academies didn't keep any one teacher on for more than a year or two, and the students came and went even faster. They did, however, establish education beyond the basic for a much larger percentage of the population than had ever attended Latin grammar schools or received private tutoring.

The real competition for most of the academies was not Harvard but the plow. Youths between eight and twenty attended them whenever there was nothing more pressing to do. The prime time for them was midwinter, when farm chores were at a minimum, though this was the most difficult time of year to get to the school. Thus, the boys often boarded in houses near the academy. It was a form of schooling that seems haphazard compared with contemporary high schools.

Still, the academies did have some advantages. They did not confuse education with custodianship of youth. The young males who attended were there for a limited, specific purpose—usually the fairly modest aims of learning to write and do bookkeeping. The academies did not try to replace the family, the church, or any other institutions of society. Academies were there for students when they needed them. Unlike the high schools of a century later, they did not organize their students' social lives, monitor their health, or strive to instill civic virtue. A male in his teens in late-eighteenth-century America typically had regular daily contact with different sorts of people. Time spent with his age-mates in school was only one aspect of a varied life.

Still, the academies treated students in their teens more as boys than as men when it came to discipline. Although educators at the time said persuasion was the most effective tool to shape the behavior of the young, the teacher's job was also to fight sin. As we now recognize certain psychological problems as beyond the range of reason, so then was Satan seen to exert a power that the teacher had to confront in the classroom. The goal was to break the student's will, if necessary by force. Physical discipline was commonplace, and teenagers were hardly exempt. Schoolmasters seem not to have distinguished between students of different levels of maturity in meting out punishment. Indeed, the greater size and power of the older students made it even more important that they knew who was boss. Teachers often subjected them to harsher and more humiliating punishments than those given younger students. It was not uncommon, though, for a youth who was treated as a child at the academy to break up his studies by working as a teacher at a district school. There, some of the students he had to discipline would likely be his contemporaries.

There are many recorded incidents of hostilities between academy students and their schoolmasters. Some of them seem to have been largely ritualistic. For example, students often conspired to "put out" the schoolmaster by barricading the school building and physically barring the teacher from entering. The earliest instances of this practice probably involved real grievances and sincere insurrection. But it quickly turned into something students were expected to do and the masters expected to happen periodically. They played along with the role reversal, fighting against but tacitly acknowledging the power of their students.

The most serious incidents of student unrest happened at colleges, which, unlike academies, did create segregated communities of the young. In most cases, the result was, basically, mischief. In 1774 Philip Fithian, a recent College of New Jersey (Princeton) graduate, looked back with pleasure on "Meeting & Shoving in the dark entries; knocking at Doors and going off without entering; Strewing the entries in the night with greasy Feathers; freezing the Bell; Ringing it at late Hours of the Night."

Sometimes the jokes shaded into rebellion. Harvard experienced several "bread and butter riots" in 1766, but things were far worse at Yale. There, a prolonged insurrection by the students, including physical attacks on several deeply unpopular members of the faculty, essentially crippled the college from 1761 to 1766. Other colleges suffered from sporadic incidents as well. The 1760s was one of the most turbulent decades on campus in our history.

And the protesters, unlike their counterparts two hundred years later, did help make a revolution.

Indeed, by 1769, Harvard, located near the center of the most militant agitation against English rule, had become a highly politicized place. "They have imbibed the Spirit of the times," reported the college's secretary. "Their declamations and forensic disputes breathe the Spirit of Liberty. This has always been encouraged, but they have sometimes wrought themselves up to such a pitch of Enthusiasm it has been difficult to keep them within due bounds." He added that their tutors did not want to destroy "a disposition which may hereafter fill the Country with Patriots and choose to leave it to age and experience to correct their ardor."

More than two centuries after the fact, the revolution is most often appreciated as a triumph of the Age of Enlightenment. The founders are seen as men of dazzling insight and formidable learning who were able to craft both the idea of a nation and an enduring system for governing it.

While this view may not be wrong, it's still worth bearing in mind the youthfulness of the country at the time of the revolution and the many roles young people played in it. We can argue about whether the revolution could have happened in a country with an older population. We cannot, however, escape the truth that the revolution happened when we had, by percentage, more young people than at any time in our history.

Young people were everywhere in the revolution, not simply as fifers and drummers but as fighters. They were part of the underground organizations that were active during the decade before fighting broke out. The revolutionaries' ships had crews "manned" by fourteen-year-olds and even younger boys who had lied about their age. Apprentices used the occasion to desert their masters. And though militias did not call up boys, they rarely turned away anyone who looked large enough to be useful.

In the sense that young people end up bearing the brunt of the death and suffering in nearly all wars, the American Revolution was typical. Unlike many other wars, though, the cause seemed to be an expression of youthful values. Certainly by the time the war was over, young people were among the most enthusiastic republicans of all.

Take, for example, one of the most celebrated incidents leading up to the revolution, the so-called Boston Massacre of March 1770. While it was the culmination of resentment felt by many radical Bostonians of all ages over the English military presence in their city, it was an affair largely of the young. It followed by a few days an incident in which a rock-throwing boy of eleven was shot dead by the man whose house he was targeting—an unpopular informer for the customs authorities.

The events that ended with English soldiers firing into a crowd began with a teenager, an apprentice wigmaker, who chided an English officer for not paying his master's bill. The officer, who had in fact paid for his wig, ignored the attack, but a lone sentry, a private and probably a teenager himself, answered the insult, which drew other youths into the shouting match. With the ringing of fire alarm bells, the few people near the sentry's post swelled into an ice- and snowball-throwing mob.

The English officer in charge that night was a young man who had just turned twenty, an army man since he was twelve. He went with eight of his men to save the lone sentry from the crowd. The crowd taunted the soldiers to fire, though many seemed to believe they were not authorized to do so. There are many contradictory accounts of how the firing started. A key witness was a fourteen-year-old apprentice who had watched it all from a window. He may have been bribed not to contradict the widespread belief that the British troops fired first and were unprovoked.

Five years later, one day after the first military engagements of the revolution at Lexington, in April 1775, Benjamin Russell, fourteen, went to school in Boston. When word came of the engagement, his master announced, "Boys, war has begun, school is broken up." Russell left school with some friends, and they made their way to Cambridge, where Harvard had virtually shut down and its buildings served as quarters for the revolutionary army. Russell remained and soon was serving in the very responsible job of company clerk. His father, a stonemason, received no word of what had happened to him, and he was worried. But it wasn't until August that he appeared in Cambridge, found his son, grabbed him by the shoulders, and shook him. He berated Benjamin for running away and not sending any word about his whereabouts. He convinced Benjamin's commander that his son was too young to serve in the army, and Benjamin was released to his father, who sent him out as a printer's apprentice.

It would be possible to recount thousands more stories of people in their teens who played a role in the revolution. My point is not that the American Revolution was made by teenagers. Indeed, it's virtually the opposite: that young people in their teens participated alongside people of all other ages. In a place and time where half the population was under sixteen, young people would inevitably play a major role in the central events of the era.

The Boston Massacre is a reminder of the fine line between youthful rowdyism and political action. Except for the greater firepower employed today, it has much in common with the unfocused, opportunist political resistance practiced by young people today in Africa, Central Asia, and the Middle East. Such anarchic activities become coherent only later, when they are described and understood. That we now refer to it as a massacre is a tribute to the effectiveness of the revolutionary propagandists.

The case of Benjamin Russell reveals several other things. He turned, virtually overnight, from a schoolboy into a man of some responsibility. This is, in part, a tribute to his own abilities, and also a reflection of people's willingness in times of chaos and stress to accept help from wherever they can get it. But the story also gives a glimpse of a society in which age was not crucial to how people were judged.

The story also shows that family obligations still counted for a lot. In the eyes of the commander, rebelling against a king was necessary, but rejecting a father was not. Young Russell may have behaved as a man, but he was obliged to show the obedience of a son.

Russell later became a wealthy and powerful man, a subject for a biographer who told this story. He was not a "typical" youth of the revolutionary era. Most young men his age were on farms and had only limited exposure to the events that were unfolding in Boston, Philadelphia, or New York. Nor was he an embodiment of the way in which life was changing for young people. He does, however, show us something that's true of most individuals in most times. Our lives look both back and forward. The choices that we make depend on a unique complex of individual circumstances. Only in retrospect can we truly give a shape to one person's story, let alone to the countless personal decisions and accommodations that constitute history.

The story that makes most sense when looking at the revolutionary period is of the liberation of youth—both by themselves and by their parents. Yet nearly any individual case you consider contains ambiguity and paradox. That's partly because the values people profess are not always those they follow, while they may not be honest, even with themselves, about the reasons for their actions. Moreover, the same action can be open to multiple and contradictory interpretations.

For example, the dilemma of Jemima Condict, nineteen, of Essex County, New Jersey, has been anthologized to demonstrate the increasing freedom of a young American woman on the eve of the revolution. In a series of letters written in 1774–75, the young woman agonized over whether she should marry a suitor who was also her cousin. She had received much advice, solicited and not. "They tell me they don't think it is a right thing and it is forbid," she wrote. "But none of them, so far as I can find out, tell me where it is forbid." (White families in colonial times did not observe any taboo against marrying first cousins, though most slaves had brought such a prohibition with them from Africa.)

Jemima decided to lay her quandary before her mother. "She said she did not think it was right," Jemima wrote of her mother's advice, "except I thought so myself." Thus, her mother advised that she form her own judgment of the legality and morality of the marriage. In the end, she didn't marry him.

One thing that is clear from this story is that arranged marriage, which would have been routine a generation before, was no longer an issue in the circle of this young woman. Moreover, she seems to have confided in many people before she took the question to her mother, let alone her father, who would traditionally have had the final say. Still, deciding to leave the matter up to her mother was not necessarily the most liberated thing a young person might have done. Indeed, the mother's endorsement of her daughter's judgment seems more revolutionary.

Jemima, at nineteen, was not very young in the context of her day. One 1781 female observer of fashionable New York parties noted that the teeth of young women over eighteen appeared visibly decayed. It's easy to forget how routine practices of modern life—such as dentistry—have a youth-prolonging effect.

Jemima was relatively well-to-do, if not rich, and appears to have been part of a trend toward somewhat later marriage among the wealthy. A generation before, when parents negotiated marriages for their children with a dynastic eye for the preservation and enhancement of wealth, young people with money married earlier. Young people without money or land to bring to the marriage had to wait until they had accumulated enough to make a marriage settlement.

As control of marriage moved from parents to the young people themselves, the age of marriage for all classes converged. The most common marriage age became twenty-three or twenty-four for young men and eighteen to twenty for young women. Those who were less well off married earlier than they had before, perhaps because the cities and the frontier were generating opportunities that made family money less important. Meanwhile, those who were wealthier began to marry later, making room in their lives for courtship rituals and occasions, such as balls, card parties, and theatrical performances where they were enacted. Such institutionalized social events as the Assembly in Philadelphia, which were in a sense public, though limited to a few, were the beginnings of what was to become known as society.

It's arguable that the privileged young women who participated in this life were pioneers of teenhood as we know it. They had few household responsibilities. Their role in life was to be pretty and charming. Young women were poorly schooled, even at the primary level, where they were typically taught only to read, while their brothers learned to write and do arithmetic.

Girls' secondary schools did not have the practical, come-and-go quality of the academies and were really closer in spirit to the traditional grammar schools. They were, in Franklin's terms, concerned more with the ornamental than with the useful. That's because the things considered "useful" for women were mostly skills like sewing that were mastered in the home. Some women had to manage complex households, including children, apprentices, servants, and slaves, but this was not seen as a subject for schooling.

It seems obvious today that secondary education for women was an important piece of social progress from which individual women and the nation as a whole benefited. Unlike most schooling for young men at the time, there was little sense of what it should accomplish. Paths to higher education and to the professions—the principal goals of men's grammar school education—remained closed. Women's schooling served an urbanizing country by providing upper-class young men with more cultivated wives. Because the friends girls made in school often lasted for a lifetime and were often replicated by their daughters, women's schooling helped to create a network of relationships among upper-class families throughout the colonies. Still, one important reason families schooled their daughters was simply to find something for them to do.

The emergence of exclusive social rituals, held in the newly built ballrooms of showy mansions, doesn't seem to support the case for the revolutionary spirit of the young. Indeed, during the revolution, when New York and Philadelphia were occupied by British troops, they experienced the most glittering social seasons they had yet known. The soldiers, with their handsome red coats and their experience in a more cosmopolitan world, impressed the young women at many parties thrown in their honor. For some socially minded young women, their compatriots' victory over the dashing British troops came as a disappointment.

Still, in another sense, the emergence in the cities of a social life was part of the liberation of the young, and especially of young women. The girl from a moneyed family was no longer a chattel to be bargained over. She was more like a gem worthy of a setting where she could sparkle. This might seem to be scant progress by modern standards. For women at the time, though, it must have felt like a great advance. They were more or less in control of this new social setting and of their own courtship. At least some people felt they had altogether too much freedom in their relationships with men. "The misses, if they have a favorite swain, frequently decline playing [cards] for the pleasure of making love," wrote the 1781 partygoer, "for, to all appearances, tis the ladies and not the gentlemen who show a preference nowadays." To a disapproving spectator like this one, such behavior was truly revolutionary.

SIX
Young Americans

But as soon as the young American begins to approach man's state, the reins of filial obedience are daily slackened. Master of his thoughts, he soon becomes responsible for his own behavior. In America there is, in truth, no adolescence. At the close of boyhood, he is a man and begins to trace out his own path.

—ALEXIS DE TOCQUEVILLE, *Democracy in America* (1835)

Here we are in 1835, six decades before many historians believe that adolescence was invented, and an extraordinarily acute observer of culture has already pronounced it dead. What's going on?

One answer is, of course, that the adolescence that emerged during the nineteenth century was very different from that which Tocqueville's upper-class French readers might have known. For them, adolescence was an age-old prerogative of a privileged class—a time of preparation and waiting before exercising the political, economic, and social power that derived from their family connections.

American adolescence, which was beginning to emerge in its distinctive form at the time Tocqueville was writing, was a luxury that relatively few families could afford for their children. It was a privilege, to be sure, but not an aristocratic one. It resulted not from any sense of the permanent position of one's family in the society at large, but rather from its opposite. The successful families of the early republic began to limit the size of their families and concentrate on their schooling and preparation precisely because they knew that their status was in jeopardy. American adolescence grew from insecurity, a belief that the only way to consolidate the gains of the current generation is to increase the knowledge and choices available to the next. It was the result of revolutionary changes in technology and organization, in the concentration and deployment of capital, and, Tocqueville would no doubt add, democracy.

Adolescence is, by definition, a dependent state, experienced in one's own household and in institutions intended for people in their teens. As it developed in nineteenth-century America, however, adolescence did require a measure of at least psychological liberation by the parents. They had to understand that their children would be, in some important ways, different from themselves, and to commit their money and attention and their children's time (and the money they might have made) to their preparation.

Another explanation for Tocqueville's premature burial of the adolescent is that he was looking back at the early republic and the emerging frontier, while ignoring the industrial culture that was, here and there, beginning to take shape. And it is true that, as we trace the various rises and falls of the fortunes of Americans in their teens, the first years of the nineteenth century do seem to mark a particularly high point.

The newly created country was generating new opportunities for its young. In 1803 the Louisiana Purchase more than doubled the country's size. The growing cities—in the interior as well as on the coast—were generating new ways to make a living. The "mysteries" of crafts, professions, and businesses were being collected in books. A literate young person could learn more from a book than from a master, and was thus freed from the dependency inherent in the apprentice system.

Still, the biggest reason that this appears to have been a great time to be young is that people in their teens were not part of an isolated group. They weren't seen as the property of their fathers, as they had been before, or as the immature, unreliable creatures who emerged in public consciousness later in the century. They weren't children, or adolescents, or teenagers.

They were young Americans, who were fully a part of the new democratic society they were rather self-consciously beginning to create. The young men were among the most eager participants in the militias found in nearly every village. Their regular drills—and especially those held around the Fourth of July—were marked by firing muskets while drinking vast quantities of alcohol. Accidental shootings were common. And there was always a party afterward. These activities may have done little to improve military preparedness, but they helped to create a social life for men of all ages.

Other sorts of community activities, such as barn-raising and church-building, also included men and women of all ages. But they were particularly prized by young men as settings to show off their strength and prowess to the entire community, and especially to young women. Similarly, young women engaged in quilting bees and other cooperative activities that integrated them into the community, and at the same time served as a showcase for their domestic skills.

And parallel to this tavern-centered secular society, the churches continued to be receptive to young men and women and make them full members of their congregations.

In fact, the first quarter of the nineteenth century brought the outpouring of religious fervor known as the Second Great Awakening. A portion of upstate New York became known as the Burnt Over District because of the religious revivals that moved through the area like wildfires, and the new sects—including Mormonism—that arose. Young women, who were increasingly taking jobs outside their homes, were among the most zealous converts.

Even though young men in their teens were not allowed to vote in most localities, they formed an important part of the membership of political clubs, reform movements, and other partisan activities. Some considered it a civic duty, and perhaps an entertainment, to attend court trials.

Many towns also offered lyceum lectures, dancing and music schools, and other opportunities for young people bent on self-improvement. Much of the information we have of such people comes from journals and diaries, which are more likely to be kept by those who are ambitious and introspective than by those who live out the expectations of others. Still, these documents tell us that a significant number of young people had the perception that things were changing rapidly, that they would be part of the changes, and that they had better prepare for them.

As always, the rewards of this republic of the young were not equally distributed. Those who had it best were young white men who either lived in towns or cities, or who had ambitions on the frontier. Not benefiting at all were slaves, whose humanity was given only slightly more legal recognition in the new republic than during colonial rule. (In Virginia, for example, the killing of a slave was redefined as murder, not mere property damage.) And their family ties were not being loosened by the law, but were rather being viewed as nonexistent. Slave families were broken up routinely when individuals were bought and sold. Some, such as Frederick Douglass, claimed not to have been hindered by the loss of their families, while others had memories of heartbreak.

For many slaves at this time, the beginning of their teens was the first moment that they fully understood the meaning of their situation. Slave children on large plantations were expected to do chores for their family, including tending the plots and animals the slave families kept, and to do minor work for the plantation. But not until the age of thirteen or fourteen did slave training begin in earnest. "It is policy," wrote one slaveowner, "to leave the slaves to grow and strengthen, unfatigued by labor, until they are old enough to be constantly occupied, as a colt is left unshackled, with free range of the pastures, until the 'break-in' time comes."

For some young women, the coming of physical maturity brought the shocking realization that even her sexuality and reproductive abilities belonged to her owner.

"After I been at de place 'bout a year, de massa come to me and say, 'You gwine live with Rufus in dat cabin over yonder,' " a former Texas slave recalled in an interview done by the Works Project Administration during the 1930s. "I's 'bout sixteen year old and has no larnin', and I's jus' igno'mus chile. I's thought dat him mean for me to tend de cabin for Rufus and some other niggers. Well dat am start de pestigation for me. . . . I don't like dat Rufus, 'cause he a bully. . . . We'ums had supper den I goes here and dere talkin', till I's ready for sleep and den I gits in de bunk. After I's in, dat nigger come and crawl in de bunk with me, 'fore I knows it. . . . De next day I goes to de missy and tells her what Rufus wants and missy say dat am de massa's wishes. She say, 'You's am de portly gal and Rufus am de portly man. De massa wants you-uns to bring forth portly chillen.' "

She resisted one more time, using a hot poker to fight off Rufus's advances. The next day the master told her he'd paid big money for her, and that she'd have children with Rufus or else be whipped. "I thinks 'bout massa buyin' me offen de block and savin' me from bein' separated from my folks, and 'bout bein' whipped at the stake. Dere it am. What am I's to do. So I decides to do as de massa wish and so I yields."

Thus, slave youths became full-grown, economically useful, and marketable at the same moment. Moreover, at least some slave narratives bolster psychological studies that suggest that this is also the period when people develop a political consciousness, an awareness of one's power and status within the society as a whole. For slave youths, the teens were not a time of transition but of profound, shocking insight, and the beginning of hard labor that would last all the rest of their lives.

Moreover, the vast majority of slaves were engaged in agriculture, by far the largest part of the economy and also the least dynamic. Even for many white youths in the North, the demands of farming and one's duty to help the family survive meant they had to delay or forgo more modern ambitions. Slaves didn't even have theoretical choices. Free black citizens in the North tended to cluster in the cities and large towns, but their opportunities were far fewer than their white counterparts.

"Before she has completely left childhood behind," wrote Tocqueville of the American girl, "she already thinks for herself, speaks freely and acts on her own. All doings of the world are ever plain for her to see. Seldom does an American girl, whatever her age, suffer from shyness or childish ignorance. She, like the European girl, wants to please, but she knows exactly what it costs. She may avoid evil, but at least she knows what it is; her morals are pure rather than her mind chaste." Unlike privileged European girls, who could be protected by her family, Tocqueville argued, democratic circumstances required that the American girl had to be able to protect herself.

This is the first description I've encountered of the all-American girl—fresh, resourceful, tough-minded, with an aggressiveness that delights rather than intimidates. Tocqueville claims to have been "almost frightened" by the skill and "happy audacity" of the girls he encountered. Although he feared that such women would lose their imagination and become cold, he was obviously charmed.

The all-American girl has been a persistent image of our culture ever since. Even though she doesn't have any current pop culture embodiment as powerful as Katharine Hepburn, or Judy Garland, or even Doris Day, she is still starring in countless books written for preteenage girls. And amazingly, after more than half a century, cute, feisty Betty is still doing battle in the newspaper comic pages with glamorous, pampered Veronica for that dubious prize, Archie.

For young women in their teens, the first half of the nineteenth century brought enormous changes. They were markedly more independent of their parents. They were taking a leading role in religious revivals and participating in all-female conferences and informal sisterhoods that linked them with other women, young and old. Near universal female literacy made them a significant market for books, and some young women were able to realize their literary ambitions. It became possible, finally, to forgo marriage. Becoming a spinster was not what young women were encouraged to do, and it may not have been what some of the women who lived single lives truly wanted. And literally being a spinster—a woman who supported herself at home with her spinning wheel—was rapidly being rendered obsolete by the mechanization of the textile industry. A single life for an American woman was not easy or very desirable, but it was no longer unthinkable.

Young women formed the vanguard of the American industrial workforce, a wholly unprecedented experience. Never before in history had young women been brought together in such numbers. Daughters had traditionally passed directly from the authority of their fathers to their husbands. Now, for many, there was an interlude between the two, living among other girls (though under the control of greedy industrialists and with little time of their own).

Most women's lives were still a reflection of their husbands'. They fulfilled the age-old function of wives, which is unspecific but comprehensive: to hold the household together. And the challenges faced by American wives could tax the strength that Tocqueville and other foreign observers saw in the young women of the time. She might find herself living by herself for long periods of time in a wilderness cabin, caring for children and fending off wild animals and sometimes hostile Indians. She might be charged with creating an elegant urban household for her merchant husband, or keeping a humble house in a factory town where her husband and her children went out to work.

The mention of a factory town and working children ought to jolt us out of our contemplation of a republican golden age. It is nice to think that there was a golden moment of community and freedom (albeit tarnished badly by slavery) before the forces that shaped nineteenth-century life fully took hold. Tocqueville certainly suggests as much, and that might be one reason for his enduring appeal. It is, moreover, possible to find writings by young people who were thrilled by the possibilities they saw in their new nation and their own lives.

Yet, each of these can be balanced by an account from a young person who is confused or depressed, who is suffering from religious doubts, or who is complaining of a tyrannical father. As the historian Harvey Graff has observed, there is rarely one path for young people to follow, but many conflicting ones. And the early nineteenth century, precisely because it seemed to offer such a wide array of opportunities, was a time of great anxiety for many who had to choose how to spend their lives.

Although this may not have been a happy time for all who lived through it, it is a poignant one for those who look back. The young were accepted as more or less full members of a society in which community values were, briefly, very strong. But as the century went on, the society of independent artisans and merchants gave way to one in which increasing numbers of people were employees. They had moved from a society oppressed by family obligations to one oppressed by the demands of industrial organization, machinery, and capital. The young women who became the first generation of factory workers literally had to choose either never-ending toil for their fathers or the machines, which demanded thirteen hours a day. Young people might have been treated much the same as their elders, but often that was very poorly indeed. Youths had attained something close to the status of adults, but much of the freedom that had been previously promised was disappearing.

The old "mysteries" of the artisan had been revealed in books for all to see. But by the time that happened, many of these crafts were already dying, to be supplanted by a new means of production with mysteries of its own. These new mysteries involved technology, industrial organization, finance, and even public relations. Compared to these new capitalist mysteries, those that had been sought by apprentices were unsophisticated and widely accessible.

As in England, the first wave of American industrialization involved textiles, a traditionally female activity. Spinning and weaving by women in the home had been for many a significant source of family income.

The machinery that took textile production out of the house and into vast mills was, at the beginning of the nineteenth century, quite literally a mystery. Neither the machines nor their plans were available for export, and only by several cases of industrial espionage did mechanized textile production come to these shores. The most important, though not the first, such case involved Francis Cabot Lowell, who was able to observe not only the newest innovation—the power loom—but the entire system of integrating machinery to provide efficient, high-volume work flow. Back in Waltham, Massachusetts, in 1810, he "reinvented" the process by copying the English machinery, and, for the first time, putting it all under one roof.

Lowell and his partners also added an innovation that was not technological but social. Rather than hire entire families—father, children, unmarried sisters and brothers—to work for a package rate, as mills in England and some in New England were doing, they reached out to a new source of labor. These were the farmers' daughters, the girls who had been spinning and weaving in the home. Work in the factories was laborious, but not beyond the capabilities of physically mature young women. "Factory girls," as they were known—young women from sixteen to their mid-twenties—became America's first and most celebrated industrial workforce.

After Lowell died in 1817, what had become known as the Waltham system was realized in its purest, most famous, and most extensive form along the Merrimack River in northeast Massachusetts in the town that was in 1822 named for him, and which became, by the 1840s, the largest manufacturing center in the country. (The population was 30,000, three quarters of them female.) Other industries operated mostly by young women soon sprung up elsewhere in Massachusetts, New Hampshire, New York, and as far west as Pittsburgh. By 1831, according to Patrick Jackson, one of the partners of the enterprise that built Lowell, 39,000 young women were working in cotton manufacture in the United States, earning wages of more than $4 million a year.

The era of the factory girls was quite short, probably less than twenty-five years, ending in about 1850. Part of a perennial American quest for "clean industry," it was an attempt to manufacture on a large scale without brutalizing the workers. The promise the mill owners made was that they would treat these native-born farmers' daughters with close to the same respect they would accord their own daughters. The reality of factory girls' lives may not have lived up to the rhetoric. Still, Lowell and his partners seemed determined, at least at the outset, to create a factory system that preserved democratic values. And despite the additional cost of maintaining the system of boardinghouses and opportunities for improvement, the system proved, for a long time, to be competitive with other factories.

Use of a young female workforce had several advantages for the mill owners. Industrialization was already associated with poverty and dependence in England, and Americans feared that it would be a blight not just on towns and families but on democracy itself. By hiring young women for what was presumed to be only a few years before marriage, the mill owners allayed fears of the creation of a permanent factory class. The young female workers also had a place to go home to if they became sick or simply disenchanted, and there were plenty more farm girls who were willing to take their place. They were considered extremely unlikely to agitate or make any trouble. And because the factory girls' income was often used not for their own benefit, but to pay the mortgage on the farm or to send a brother to college, the majority of Americans who lived on farms had an interest in the success of industry.

"Daughters are now emphatically a blessing to the farmer," Jackson declared. The blessing of which the mill owner spoke was not religious or sentimental but economic. For the first time, women's value in the home could be balanced against what she could receive outside it.

This newly emergent female earning power helped overcome the general prejudice against factory work. Parents today justify their teenage children's employment as a character-building exercise, though their children will tell you that they're doing it for the money. The mill owners of a century and a half ago similarly sought to persuade parents and young women that working in the mills was, at its heart, an opportunity for self-improvement.

This approach imposed some burdens on the mill owners. They had to be able to prove to the girls' families and to residents of the towns where they operated that the girls were living virtuous lives. In practice, this meant that the girls were required to live in boardinghouses within a short distance of the factory gates, most of them run by widows who received a stipend from the manufacturers as well as rent from their tenants. The girls were also required to attend church services on Sundays, their only day off. (Some girls found this requirement not only intrusive on their meager leisure but expensive; they had to purchase special dresses for church and spend time washing and ironing them as well.)

And unlike other mill owners, who often paid only in scrip redeemable at a company store and drew their workers into a web of debt and dependency, those who hired girls had to pay cash. That's because the young workers were sending most of their wages home. Moreover, at least at the outset, the wages were quite attractive, in order to overcome the stigma of factory work while making more respectable callings, such as teaching school, unattractive by comparison. It's worth noting, though, that while the salaries paid to factory girls were higher than those available to women anywhere else in the country, they were far lower than those paid to able-bodied men.

> *We go in at five o'clock; at seven we come out to breakfast; at half-past seven we return to our work and stay until half-past twelve. At one, or quarter-past one, four months of the year, we return to our work and stay until seven at night. Then the evening is our own, which is more than some laboring girls can say, who think nothing is more tedious than a factory life.*
>
> —*"SUSAN," quoted in* The Lowell Offering *(June 1844)*

That sounds like a hellishly difficult day. Moreover, consider that some of the boardinghouses were as much as a quarter mile from the factory gates, and that only half an hour was allotted for the round trip and the meal itself. And yet, as the writer hints, there were plenty of girls who felt that they hadn't a moment they could call their own and were thus susceptible to "Lowell fever."

Indeed, much of what was written in favor of Lowell and work in the factories was, in fact, an indictment of the laborious monotony of life on the farm and an expression of the desire of young women to be able to have some time of their own, to have some money to spend, and to escape from tyrannical parents and stepparents they despised. "Susan," in the series of letters quoted above, advises her friends back home on whether to come to Lowell. Those who are simply curious or restless but otherwise well-off she tells to stay home. Those who are miserable, hopeless, or furious she advises to come.

A more famous theme of the considerable literature spawned by Lowell concerned the opportunities for learning, literary expression, and cultural improvement available to the girls. Biweekly lectures by some of the most eminent thinkers of the day alternated with concerts. Girls took night courses in everything from botany and German to phrenology. Girls organized self-improvement circles to read and discuss current issues and improve their writing skills.

There was a mania for reading among young women during the late eighteenth and early nineteenth centuries, and novels featuring young women were among the bestsellers both in England and the United States. Parents worried that so much reading was a kind of disease, much as later generations of parents have worried about their children's enthusiasm for movies, jazz, television, rock and roll, and the Internet. Girls' diaries often express guilt about their excessive novel reading and resolve to change their habits. "Books of instruction," one Virginia sixteen-year-old assured herself, "will be a thousand times more improving."

Lowell boasted several very popular libraries, and some girls are said to have come simply because there was nothing to read in rural Maine or New Hampshire. While all the girls had received common schooling and illiteracy was virtually unknown, most came from places where they had scant hope of further education. These opportunities did exist in Lowell, and there were girls who took advantage of them, though probably not as many as had intended to. Not surprisingly, the frantic fourteen hours from first bell to last made them too tired to accomplish much during the remaining ten hours of the day.

There were, however, some advantages to getting out of the boardinghouse for the couple of hours after supper and before the front door was locked and lights extinguished at 10 P.M. For one thing, the houses were very cramped, with girls sleeping two and sometimes three to a bed, in dormitories of six to twelve girls.

Those who remained in the ground-floor common room frequently found themselves besieged by salesmen offering candy, books, clothing, shoes, utopian theories, and fortune-telling. A writer in *The Lowell Offering* deplored their "breaking in upon the only hours of leisure we can call our own, and proffering their articles with a pertinacity which will admit of no denial." But she added, regretfully, that some of the girls seemed to enjoy the attention. For the first time ever, there were large groups of young women, all of them with at least a little money of their own, all in one place. In the boardinghouses of Lowell, we can glimpse an early version of that consuming powerhouse—the youth market.

Factory girls, and particularly those working at Lowell, held a special fascination to observers of the early nineteenth-century American scene. The English novelist Anthony Trollope, who described Lowell as "a philanthropic manufacturing college," visited a boardinghouse and marveled that the girls ate meat twice a day—and that it was hot. President Andrew Jackson watched 2,000 of them parade past in identical white muslin dresses and pronounced them "Very pretty women, by God!" Several commentators viewed the factory girls as nuns of industry, and editorialists marveled that, while in England, industry was dehumanizing, in America it had produced what amounted to a new privileged class. "No where else does a Laboring class of equal numbers earn so much, year by year," wrote the *Scientific American* in 1845, "no where else are they so constantly employed, comfortably situated, and adequately rewarded."

One eminent dissenter was poet John Greenleaf Whittier, who lectured regularly at Lowell and lived there for a time. He was one of the first celebrators of the place, but he changed his mind. "Every web which falls from these restless looms," he wrote, "has a history more or less connected with sin and suffering, beginning with slavery and ending with overwork and premature death."

The *Scientific American* article was, in part, a reaction to increasing labor militancy at Lowell and the other factory towns operated primarily by young female labor. During the 1830s in Lowell, there had been two "turnouts," or strikes, in response to the mills' reduction of wage rates. Perhaps the most dramatic of the turnouts occurred in Pittsburgh in 1845, an event that lasted several days and involved sporadic street rioting. Suddenly, in accounts of the time, these ethereal industrial maidens metamorphosed into amazons, seeking to overturn the natural order of things. In 1845 a Massachusetts legislator from Lowell thought he could turn aside women's petitions for a ten-hour day by forcing them to testify in public, something he thought they'd be too modest to do. But they did testify, and then they organized and succeeded in defeating him in the next election, which was quite an achievement for people who couldn't vote themselves.

After each outburst of militancy, though, the vast majority of the women returned to their machines at the lower rates. Most increased the amount of work they did, tending three or four looms, for example, rather than the two that had been standard, so that they could keep their pay at the same level. Their work seems not to have been so voluntary as workers assumed. Others were depending on them. Still, it's probably true that no group of women anywhere had ever before received so much cash for what they did.

One reason that the factory girls still command our attention today is that a few of them wrote. For several years, some of the girls of Lowell produced a magazine called *The Lowell Offering*, a publication that was hailed throughout the world and, undoubtedly, read more outside the city than in it. On two continents, those who perused it expressed their amazement that young women working in cotton factories actually had minds and were able to express themselves. At a time when many in the professional and literary class were coming to believe that there was a "natural," virtually subhuman working class, *The Lowell Offering* suggested that enlightened industry could allow workers to keep body and soul together. That the writers were young women made a further point.

There's no question that the magazine enhanced the image of the mill owners and helped insulate them against arguments that their work-day was longer than the satanic mills of England that Americans so abhorred. The *Offering* did not speak for the interests of factory girls as workers, nor did it recognize the nascent labor movement that was organizing in some of the mill towns.

The best of what appeared in the *Offering* expresses a clear-eyed, generally unsentimental look at a new way of life. These were young women who had to make some difficult calculations about their lives under unprecedented circumstances. Young women making choices, and writing about how they did so, was an exciting phenomenon, even if they stopped short of analyzing and trying to change their situation. Some stories are filled with details on how different sorts of girls choose to spend the precious bit of their wages they were allowed to keep, and their even more precious time.

In one, entitled "The Night Before Payday," we hear girls chatter about how they'll spend their wages once "the beautiful paymaster will come in, with his coppers rattling so nicely."

> *"Well," said Elizabeth, "I shall sow my wild oats now, and when I am an old maid I will be as steady, though* not quite *as stingy as Dorcas. I will get a bank book and trot down Merrimack Street as often as she does, and everybody will say, 'What a remarkable change in Elizabeth Waters! She used to spend all her wages as fast as they were paid her, but now she puts them in the bank. She will have quite a fortune for someone, and I have no doubt she will get married for what she* has, *if not for what she* is.' *But I cannot begin now."*

Many women reading that story in 1841 would have envied the freedom of these women, the confidence of their talk, and even the camaraderie of the boardinghouse. Many of the articles express contempt for wealthy girls, who have time for schooling and reading and money for the finest clothes, and who look down upon those working hard to gain the same things. For the many poorer women on farms who were tempted by Lowell, the mere fact that one could work and acquire some of the trimmings of fashion was a novelty.

Much less well known than the *Offering* was another paper published in Lowell, *The Voice of Labor*. Several of Lowell's female workers had a hand in putting it out, and Sarah Bagley, whose work had been rejected by the *Offering*, was, for a time, its editor. "Now to me," she wrote in an 1846 article, "it seems somewhat contradictory to hear those who contend, long and loud, that we have a 'moral police,' so vigilant that it is hardly possible for an operative to be vicious. . . . Think you the *benevolence* of the 'powers that be' ordained the 'all-day system' of labor? Was it not rather their avarice?"

She had a point and she expressed it well. Yet, the polemics of Bagley and her militant cohorts sometimes seem less feminist than many of the stories in the *Offering*. The militants frequently argued that factory girls would be ill-prepared to become wives and mothers, and they presented a rather idealized view of domesticity. As a tract published in 1845 by the Lowell Female Reform Association expressed it, "Quiet, neatness, and calm serenity should sanctify and render almost like heaven the home of domestic union and love!" In fact, the writer was not a reactionary; she went on the call for full female equality. But unlike the writers for the *Offering*, who looked little beyond themselves, and whose nostalgia was more for nature and particular family members than for an ideal home, the militants were looking at society more broadly. Along with the factories, rapid urbanization, and increasing immigration, a new idea of home was taking shape. It was not, as it had been, a unit of the economy, but a spiritual haven, a refuge from the productive world.

The tract also argued that the girls began at the factory "as mere children." Only a few years before, sixteen had been considered grown and virtually adult. And already, many children as young as seven and eight were laboring in factories, under contracts that committed entire families to work for below-subsistence wages, often to pay off debts to the company store.

Still, the tract and other writings like it show that even the most militant of the factory girls were beginning to accept a new vision of the middle-class family. Before industrialization, nearly all homes—both in town and country—had been centers of economic production. This remained true on the farm, but in the city, the home was increasingly seen as a refuge from the world of work. And mothers and daughters took on a new role. They were apart from and, in a moral sense, above the family economy. Just as the home represented an alternative realm to that of the workplace, women's values, manners, and learning made them exemplars of higher things.

Louisa May Alcott's *Little Women* is an enduring work of this era that expresses the pull many young women felt toward greater independence, on the one hand, and a newly enhanced role as the moral and intellectual force of the household, on the other. The four March daughters in the book follow four contrasting ways of becoming young women. Jo, the boyish and literary one, is clearly closest to Alcott's heart and has probably always been the sister that readers of the book liked best. But Alcott made clear that it was the girls' loving, wise, resourceful mother who made her daughters' happiness possible. She was, by midcentury, the feminine ideal.

In many towns and cities of America, you can still see the evidence of the moment when the ideal of sheltering, female-dominated domesticity took hold. It happened during the 1840s and 1850s, as the Greek revival architectural style gave way to the Italianate style characterized by brownstone-fronted rowhouse versions of Renaissance *palazzi*. Near the cities there rose suburban villas, rural residences whose focus was family life, not agriculture and husbandry. These were not a new phenomenon on the American scene, but they had previously been restricted to the rich. Now they were a bourgeois aspiration, one that became more practical with the building of the first commuter railroads during the 1850s.

These city brownstones and suburban villas strike our eyes as remnants of a gentler era. At the time, however, this approach to family life was a response to serious crises in the lives of the merchant, artisan, and professional classes. The 1840s and 1850s brought a widespread collapse of family businesses as artisans' crafts became increasingly mechanized, and improvements in transportation and communication made it possible for wholesalers and retailers to do business not just locally but throughout an entire region.

In retrospect, we can see that what these people were experiencing was the beginning of a process that would create vastly more wealth and countless new jobs, and turn the United States into an economic superpower. But what these urban, upper-middle-class families felt was a real threat to their livelihoods, and particularly to the futures of their children. It became obvious even to many who were leading families of their communities that it would be difficult to pass that status on to their children. They were right to be concerned, and they responded in two ways that, in the long run, helped fuel the progress that was to come.

The first of these key decisions was to have smaller families. This was an intimate matter, one that was not discussed, for example, in the editorial pages of magazines. Nevertheless, the magazines' back pages had many advertisements for tonics and devices that promised, not too subtly, to induce miscarriages, or, like "Golden Female Pills," to prevent conception. Abortion, at the time a more dependable way of cutting the size of families, was also apparently widespread. The change was dramatic. A study of Utica, New York, women who entered their reproductive cycle in 1830 found that they produced 3.6 children apiece, down from 5.8 children for those twenty years older.

Statistically, this voluntarily lowered fertility of the bourgeoisie was masked by the continued high reproductive rate of the poor and immigrant populations. But it was an important step in the evolution of the teenager we know today. It focused parental energy on fewer children and helped shift teenagers of this class from a potential income source to a locus for investment.

This gave rise to the other key middle-class response to economic upheaval: an increased emphasis on education for their young by those who were able to pay for it. Those who were able to forgo the income from their children's labor were increasingly willing to invest in education so that the members of the next generation could stand on their own two feet. This did not necessarily mean traditional grammar school education leading toward college. But it did mean something more than the intermittent, catch-as-catch-can schooling offered by the academies. And it did mean that for one important segment of the population, the teen years were becoming a time of preparation, rather than the beginning of a lifetime of employment.

During the last century and a half, far more Americans have failed to embody this vision of the sheltered family than have achieved it. Yet it remains powerful. Today's working mother who cannot live up to the Victorian ideal as "angel of the household" still has occasional twinges that she ought to be doing so. Enormous amounts of our national wealth are spent moving families ever farther away from dangers—particularly those posed by other people's children.

Relatively few families in the nineteenth century were financially able to let their teenage daughters, let alone their sons, become a leisured class engaged exclusively in preparation for an adulthood that was many years off. Yet, the widespread belief that this would be a desirable goal was an important step in the rise of the idea of the teenager.

Immigrants were, of course, one of the things that middle-class families were trying to protect themselves from. Immigration was also what brought an end to the era of the factory girls. The Irish potato famine of 1846 was to prove a watershed in American life, and an end to republican ideals of social equality or industrial reform. Quickly, the major East Coast cities and mill towns were filled with hundreds of thousands of immigrants who were willing to work for low wages and who did not have to be treated with even a pretense of gentility. These new immigrants were poor and desperate, and many saw them as subhuman.

The most striking thing about the immigrants was the sheer number of them. In 1852, for example, New York received more than 300,000 immigrants, 40 percent apiece from Ireland and Germany respectively, with most of the rest from England. Despite what writers at the time seemed to believe, most did not stay in the city. But many did. The 1860 census counted 427,000 native-born people in Manhattan and the Bronx, and 386,000 foreign-born. Brooklyn was more than a third foreign-born, while Boston, Philadelphia, and the greater Pittsburgh area were more than a quarter foreign-born. Indeed, for several of these cities, the foreign-born population was greater than that of the cities as a whole only thirty years before.

In an 1847 sermon on "the dangerous classes," the liberal Boston clergyman Theodore Parker argued that nations, like boys, go through the stages of being "animal," then "savage," then "barbarous" before reaching civilized manhood. "So in the world," Parker argued, "there are inferior nations, savage, barbarous, half-civilized; some are inferior in nature, some perhaps only behind us in development—on a lower form in the great school of Providence—negroes, Indians, Mexicans, Irish, and the like." It's an interesting list. Negroes were slaves. Indians were being exterminated. In the Mexican War, the United States had just added to its territory, taking California and most of the Southwest. And in Boston, the Irish were the problem at hand, perceived by many as an alien mob given to drunkenness and crime, producing too many children, and adhering to a superstitious religion that retarded their progress.

Parker believed that adult immigrants were probably beyond hope. Only the children and the youths offered the potential of becoming good Americans. And, it was always noted, trying to save these children was not a wholly unselfish gesture. There were great numbers of them, and if they were allowed to fall into undisciplined lives of thievery, prostitution, dissoluteness, and rioting, everyone would be threatened.

The 1840s brought a new wave of public concern over juvenile delinquency. New York had violent street gangs. Those that were made up primarily of immigrants' children were the focus of the greatest attention, though there were similar gangs made up of sons of native-born, though poor parents, whose livelihoods were menaced by the arrival of immigrants.

Likewise, many observers noted a shocking rise in prostitution, involving, as one described it, "females of thirteen and fourteen walking the streets without a protector, until some pretended gentleman gives them a nod and takes their arm and escorts them to houses of assignation." In 1854, a newspaper reporter counted more than fifty young women soliciting on a mile of New York's Broadway. It's likely that the great majority of these young prostitutes were native-born. Jobs that had previously been taken by young women in their teens were those most likely to be taken by immigrants. In factories, they made way for able-bodied men, often accompanied by their whole families. And in households where they had worked as servants, young American women were displaced by Irish servants, whose greater subservience and dependence on their employers made them seem better suited for the role.

"What are you bothering me about schooling for? Education is only for them that are learning to be gentlemen." That's what Frederic Kenyon Brown, who immigrated from England as a boy during the 1850s, recalled his father telling him at age eleven. "You're a poor boy and must be thinking about getting to work. Here we are, head and ears in debt! Up to our neck in it right away. We owe for the furniture. That chair you're sitting in isn't paid for. The stove is not paid for. Nothing's ours, hardly the clothes on our backs. . . . You've got to knuckle down with a will, young man, and help us out of the hole we're in!"

As in contemporary discussions of youth crime, those concerned with juvenile delinquency in the time of which Brown wrote placed a greater emphasis on morality than on poverty—"the hole that we're in," as

Brown's father put it. Still, the things his parents set him off to do to help them stay alive frequently verged on, or crossed over, to the illegal. The most important of these was scavenging—combing the streets for chunks of coal, scraps of wood, and anything else that might be useful at home. Sometimes it wasn't clear whether such an item might belong to somebody. Sometimes youths would fight among themselves or band together to protect their booty from being stolen by bullies. It was a short, not always visible, step from scavenging to thievery and gang activity.

Those concerned with juvenile vice worked particularly hard to prevent such scavenging by girls. The best pickings, they observed, were most likely to be found near piers where the girls might be tempted into the better paying business of selling their bodies.

Brown's scavenging days were, in fact, fairly short, because his father had taken the precaution of falsifying his birthdate when he arrived in the country. Thus, he was able to leave school after only a few months and go to work at the mill before the Massachusetts legal minimum age of thirteen. In fact, Massachusetts had one of the earliest and most stringent compulsory education laws, requiring everyone under fifteen to receive at least three months of schooling each year. But there was no mechanism to enforce this law, nor was any proof of age required. Mill owners who tried to comply with the law simply lost their workers to other manufacturers who didn't bother. Educators, notably Horace Mann, the foremost advocate and architect of the public education system as we know it, deplored this laxity and sought to convince the populace that schooling would reduce crime and social unrest. This position was, however, way outside of the policy consensus of the time. The New York publisher Horace Greeley, for example, won a reputation as a child labor reformer when, in 1850, he advocated limiting the workday of children under twelve to a mere ten hours a day.

In Brown's case, his smattering of primary education came in a school intended specifically for prospective mill hands, a school whose curriculum was based on what would be useful in the mill. He later recalled that he and his classmates were enthusiastic about their impending move out of the childish world of the school into the grown-up realm of the factory. "After school," he wrote in his 1911 memoir, *Through the Mill*, "when we mixed with our less fortunate companions, who had years and years of school before them, we could not avoid having a supercilious twang in our speech when we said, 'Ah, don't you wish *you* could go into the mill in a few months and earn money like *we're* going to do?' "

Despite his working, though, Brown's family never did get out of the hole. His father drank more and more, and his mother, who at first was trying to drag him away from it, increasingly began to join him. Brown continued at the mill, despite "infinite weariness" and frequent dizziness that he ascribed to having to work so hard while his body was growing rapidly.

America had a drinking problem that long predated mass immigration. The rum-soaked revelry of the early republic was bound to induce a hangover, one that would affect the lives of the young. In the eighteenth and early nineteenth century, youths began drinking at extremely young ages by current standards—often as children, and certainly by eleven or twelve, when they began taking on adult tasks. Moreover, many medicines with high alcoholic content were administered from infancy onward. Yet, throughout the entire nineteenth century, there was hardly any public concern about drinking by the young. While contemporary Americans worry that the teenage years are the formative moment for lifelong vice, our predecessors, perhaps more rationally, saw adult irresponsibility as a blight upon the young. Still, they had a blind spot that seems painfully evident to us today: They viewed poverty purely as a result of drunkenness. They ignored the possibility that alcohol abuse was a response to hopeless circumstances, or even that mill owners encouraged drinking as a way of maintaining their workers' dependency. They did not want Brown's father to get out of the hole and be able to demand better wages and working conditions.

For many others, parental drunkenness led to patterns of abuse that spurred children of ten or eleven to leave home. Many supported themselves in the "street trades," selling newspapers, blacking boots, doing errands, while others joined the juvenile rabble of vagrants, beggars, and thieves known as "street Arabs."

Concern about the idle young, particularly in cities, actually began more than a decade before the great wave of immigration. It reflected a changing conception of the relationship of the family and the society as a whole. Previously, the common interest the society has in its children was expressed through laws and institutions that sought to strengthen the authority of the family over its members. Youth reformers created institutions, such as houses of refuge, youth farms, reformatories, industrial schools and newsboy boardinghouses, whose goal was not so much to punish wayward youth as to "Americanize" them. Many youth advocates believed by midcentury that homeless boys might be better off than those saddled with irresponsible and dependent parents—especially if the boys had opened savings accounts.

Cities tend both to concentrate social problems and make them far more visible. The products of what we'd now call dysfunctional families were on the streets of the big cities for all to see. Young people who were clearly beyond their parents' control frightened the public at large. The law generally made few age distinctions about responsibility for crime. "If acquitted," said an 1827 report of the New York Society for the Reformation of Juvenile Delinquents, "they were returned destitute to the same haunts of vice from which they had been taken, more emboldened to the commission of crime by their escape from present punishment. If convicted, they were cast into a common prison with older culprits to . . . acquire their habits and by their instruction to be made acquainted with the most artful ways of perpetrating crime."

The solution proposed was new kinds of reformatories, the first of which were built by philanthropists, then later by government, first in the major cities and later throughout the country. What these had in common was an emphasis on teaching useful skills, life in small familylike situations to provide an order that the young people's actual families presumably lacked, and release into jobs. Because many reformers saw the youths' problems to be the city itself, they sought to resettle troublesome youths on farms. They claimed great success for such efforts, though it's not really clear whether the success lay in giving the troublesome youths better lives—or simply getting them out of sight.

Unlike those who founded the juvenile court system at the turn of the twentieth century, this generation of reformers did not see the young as fundamentally different in nature from adults. They saw them simply as people who had less experience and were more likely to be set on the right course in life than those who were older. The reformers were vague about the ages of the people they sought to serve. They had to be big enough to cause trouble—at least nine or ten—but they were usually younger than sixteen, an age by which people were viewed as mostly grown, and their character formed for better or worse.

These reform efforts were under way before the Civil War, but their urgency increased after New York had experienced one of the worst urban riots in American history. It began on July 13, 1863, as a protest, mostly by young Irishmen, over conscription into the Civil War. At the time, potential conscripts could pay $300 to hire a substitute to fight for them. That was a price few Irish could afford. Moreover, the Irish had displaced free blacks from jobs as servants and other low-status workers, and some feared that if the slaves were freed, they would stream north-ward and displace the Irish from their precarious position at the bottom of the employment ladder. During the four days of rioting, an estimated 1,200 people were killed, 800 injured, and more than 100 buildings burned, including the black orphan asylum and other buildings that served black people. Eighteen black people were hanged and five drowned. At one point, a mob of 2,000 threatened to burn down the city hall, but 200 police, under the direction of a superintendent named Kennedy, held them off.

This incident tends to get lost amid the far greater carnage of the Civil War itself, which, like most wars, was fought by a rank and file made up largely of young men between fifteen and twenty. Moreover, as we look back at the Civil War, one of its particular horrors was that it was fratricidal; the young men killing each other were far more alike than not. But for the leading people of New York and other large cities, the draft riots suggested the presence of a large, very dangerous popula-tion of alien young.

According to Charles Loring Brace, founder of the New York Chil-dren's Aid Society, the majority of his city's threatening young at mid-century were American-born of foreign parents. He insisted that those he discussed in his influential book *The Dangerous Classes of New York* were fully American, but for him, that only made them more worrisome. "The intensity of the American temperament is felt in every fibre of these children of poverty and vice," Brace wrote. "Their crimes have the unrestrained and sanguinary character of a race accustomed to over-come all obstacles. They rifle a bank, where English thieves pick a pocket; they murder, where European *proletaire* cudgel or fight with fists. . . . The murder of an unoffending old man . . . is nothing to them." In seeking to horrify his readers, Brace sounds almost proud of the young people's enterprise.

Indeed, the great hope of Brace and many of his counterparts was that, either voluntarily or forced by circumstance, these young people were likely to be alienated from the parents who were the source of their problems. "The mill of American life, which grinds up so many delicate and fragile things, has its uses when it is turned on the vicious fragments of the lower strata of society."

Like many other reformers, Brace depended on the voluntary efforts of women from well-born families, whose idea of a proper upbringing was based on the newly elaborated image of the sheltered, middle-class Protestant family. The Catholic faith of most of the Irish and many of the German youths they sought to reform was something they hoped to grind away, an effort that aroused enormous resistance from families and especially from priests, who were influential members of immigrant communities. Moreover, even when they were careful to maintain a stance of religious neutrality, they were expounding values that could be sustained only by a relatively small and well-to-do segment of the population. The poor couldn't afford "normal" family life.

As in cities today, young people in their teens were not always employable, and some channeled their energy and enterprise into illegal pursuits. In the country and on the frontier, however, every available laborer could be useful. Brace's group was the most energetic exporter of urban children and young teens to the countryside. Somewhat haphazardly, it sent thousands of boys and girls off by boat and train to destinations as far away as Iowa. They were accompanied by agents who gathered farmers and tradesmen who were willing to hire the youngsters, and even take them into their families and give them land. Not all of them stayed, though Brace claimed that none turned out to be a criminal.

This passage from Horatio Alger's 1868 novel *Ragged Dick* gives a slightly different view of Brace's efforts:

> *"Has he gone?"* asked Johnny, his voice betraying anxiety.
>
> *"Who gone, I'd like to know?"*
>
> *"That man in the brown coat."*
>
> *"What of him. You ain't scared of him, are you?"*
>
> *"Yes, he got me a place once."*
>
> *"Where?"*
>
> *"Ever so far off."*
>
> *"What if he did?"*
>
> *"I ran away."*
>
> *"Didn't you like it?"*
>
> *"No, I had to get up too early. It was on a farm, and I had to get up at five to take care of the cows. I like New York best."*

"Didn't they give you enough to eat?"

"Oh, yes, plenty."

"And you had a good bed?"

"Yes."

"Then you'd better have stayed. You don't get either of them here."

Johnny could not exactly explain his feelings, but it is often the case that the young vagabond of the streets, though his food is uncertain and his bed may be any old wagon or barrel that he is lucky enough to find unoccupied when night sets in, gets so attached to his precarious but independent mode of life that he feels discontented in any other. He is accustomed to the noise and bustle and ever-varied life of the streets, and in the quiet scenes of the country misses the excitement in the midst of which he has always dwelt.

Alger's account of the young vagabonds making their living by their wits in the shadow of the wealthy people on Fifth Avenue is unquestionably a romance—even if its hero is headed inexorably for a respectable future. You can actually tell that Dick, the fourteen-year-old box-dwelling bootblack hero of Alger's novel, is going to succeed on the very first page when he affirms, "Oh, I'm a rough customer! . . . But I wouldn't steal. It's mean."

Alger wrote dozens of novels about poor, street-dwelling youths with dirty but honest faces who deserve the good luck that befalls them. Their eager readers were young people of about the same age, though presumably in more comfortable circumstances. While the values that Ragged Dick and other Alger creations embody are obviously very different from those of the rap performers who sell millions of records to today's suburban young, there is, nevertheless, a connection. In both cases, the urban youth is free of family, living by his wits and making life-and-death decisions. That's an exciting thing for those whose lives are spent preparing for some ill-defined future.

The romance of the streets comes through even in Brace's writing. After describing the hunger, violence, and physical hardship these young people suffer, he declares, "Yet, with all this, a more light-hearted youngster than the street boy is not to be found."

Alger's Ragged Dick, a proud and resourceful member of the homeless self-employed, was the sort of young man for whom Brace's organization had established newsboys' boardinghouses. Alienated from his family, he was mercifully free from their bad influences, and from the influences of Catholic priests, who were particularly skeptical of those trying to set youth on the path of enterprising Protestant values. A fiction, to be sure, though based both on reality and reformer's rhetoric, Dick was, in fact, exactly what members of the middle class hoped their children would be—a self-made man in the making.

SEVEN
Counting on the Children

In the summer the chores were grinding scythes, feeding the animals, chipping stove-wood, and carrying water up the hill from the spring on the edge of the meadow etc. Then breakfast, and to the harvest or hay-field. I was foolishly ambitious to be first in mowing and cradling, and by the time I was sixteen led all the hired men. An hour was allowed at noon for dinner and more chores. We stayed in the field until dark, then supper, and still more chores, family worship, and to bed; making altogether a hard sweaty day of about sixteen or seventeen hours. . . .

—JOHN MUIR, *My Boyhood and Youth* (1913)

John Muir, the great wilderness advocate, spent his teens helping his family subdue a piece of Wisconsin. Like many others on farms, especially in northern frontier areas, young John lived a life of punishing industry under the constant control of his father. He felt little of the freedom that Tocqueville had observed among young Americans only a generation before. Rather, he was tied closely to his family, who counted on him for its survival.

As Muir later remembered it, in 1849 he was an eleven-year-old schoolboy in Scotland, studying his Latin and algebra, when his father told him to put his books aside. He was a schoolboy no more. Soon he would be on the farm, working from before dawn till after dark, working through sickness and storms, to make the land produce a livelihood.

This kind of abrupt shift—from the life of a child to that of a worker on whose labor others depend—was one that many young people experienced in nineteenth-century America. It happened to immigrants like Muir and native-born youth as well. Although the upper classes were prolonging the immaturity of their offspring by sending them to school, most young people were seeing their childhoods shorten. And youth—

that period of semidependency when young people of earlier times were able to learn and test themselves away from their parents—was disappearing. The reason was simple: Many parents were dependent upon their children.

Almost always, this dependence was economic. Prevailing wage scales did not allow a man to make enough to support his family. Moreover, many employers preferred to hire families, and their wage scales made teenage sons particularly important for the families' livelihood.

Many parents had little choice but to depend on their children, either because the exigencies of life on the frontier and the farm, or the wage scales of mines and factories forced them to. Reformers at the time often spoke of drunken, licentious parents who sent their children off to work while they spent their days in dissipation. Still, while alcohol worsened the lot of many working families, few fathers could afford to stay home and live off their children. In the mills and mines, farms and ranches, everybody—young and old—had to work. And while there were scattered, gradually growing flurries of concern over child labor, teenage members of the working class weren't considered children.

In frontier situations, parental dependence went beyond economics to sheer survival in the face of extreme weather, Indians, wild animals, and outlaws. And in the cities, immigrants from overseas—especially those who didn't speak English—had to depend on their children to cope with American culture and teach them how to survive in a new land. Some immigrants went directly to the West, but even for the majority who didn't, the city was a frontier of sorts. Their children were scouts who were always threatening to become savages.

If you were in your teens during the second half of the nineteenth century, you would likely have been, in one sense, more "grown up" than either your immediate predecessors or contemporary teenagers. If you were not among the majority of teens on farms, you would more than likely have been working for wages at an adult job to help support the family. Yet, despite your income and responsibility, you would not have been able to feel self-reliant. More likely, you would have felt oppressed. Sons' and daughters' debts to their families had increasingly to be paid in the form of long hours of grueling, debilitating labor for the profit of strangers.

There are two distinct sorts of stories about Americans in their teens during the second half of the nineteenth century. One set describes schooling, prolonged dependence on one's parents, and the awareness of preparing for life. Only a small minority had these experiences, though they tended to be the sort of people who left memoirs, some of which will be discussed in the next chapter.

Stories from the second group are diverse in their details because working on a farm was different from a working in a mine, big cities were different from company towns, and living in the harsh weather of the northern plains was different from living in the benign climate of California. Still, they all concern hard work, started young, in order to help your family survive.

These are very different sorts of stories, but they do have a common thread. Both describe lives in which young people were losing a degree of freedom that American young people had taken for granted earlier in the century. Except in some frontier areas, the gradual liberation from the family that had earlier been accomplished by letting young people live with other families to study or work, or to apprentice to a trade, had disappeared. Suddenly, in about 1850, having young boarders in one's house began to be disreputable—something that spoke of poverty, not moral leadership as it had earlier. Apprentices rarely lived any more as members of their masters' families. And as high schools increasingly took the place of academies, students were expected to live at home, rather than to board near school. Indeed, one of the principal arguments made in favor of high schools is that they allowed children to remain at home. This was praised, in part, as a cost saving, but primarily because it afforded an opportunity for moral guidance and nurturing that only a loving home could provide.

Among rich and poor alike, then, young people were more closely tied to their families, both economically and psychologically, than they had been before.

Nearly all Americans today have grown up with the rhetoric that underlies the first set of stories. We accept the sheltering family and the close-knit, loving household as the ideal, if not the norm. Moreover, when the reality of family life falls short of this vision, we tend to view it as an aberration, or a personal failure, rather than as evidence that the vision doesn't work very well.

Thus, it requires an act of imagination to see oneself in the place of the majority of young people for whom the family was not a refuge, but rather a beleaguered economic enterprise. Of course, every family, even today, is an economic unit, and it must make calculations about who must work and how to balance domestic life with the necessity of employment. Teenagers are still part of each family's financial decisions. Do they need to work to save for college? Or should they be expected just to pay for their clothes, car, and entertainment, apparently self-indulgent expenses that nevertheless include some necessities? Nowadays, parents often work longer hours so that they can afford to provide the home environment and education they want for their children. When late-nineteenth-century families performed such calculations, the bottom line was harsh: Everyone had to work long hours so that all could afford to live miserably.

Children and teenagers have always helped their parents, though in earlier times, their aid was more likely to be on the farm or in a home-based workshop or office. Throughout the late nineteenth century, especially in cities and towns, more and more work moved from the home into large factories and commercial workplaces. The home had changed from a center of production to a site of, and spur to, consumption. This happened not only for the relatively privileged, who were able to move into sylvan neighborhoods far from the workplace, but also for the lower-paid, whose small houses were huddled near the mine or the mill. Many of these workers hired themselves out as whole families—a system that seemed to preserve a primal bond, though in practice it subjected children to terrible abuse by their employers.

The number of young people working at paying jobs rose continuously throughout the late nineteenth century. In 1870 about 13 percent of young people ten to fifteen years old—more than 19 percent of males and nearly 7 percent of females—were working at jobs. This number of employed youths doesn't seem all that high, until you consider that it doesn't include those who were working without wages within their family. That means that just about all young people on farms, where the vast majority of Americans lived, were not counted as employed, even though they were essential to their families' livelihoods.

By 1880, the percentage of young people employed had risen to about 17 percent, and the actual number of people employed in that age group had increased by more than half. By 1900, the number of young people employed off farms had risen another 50 percent and comprised more than 18 percent of the population of ten- to fifteen-year-olds—more than 26 percent of the boys and 10 percent of the girls.

In contrast, the fraction of the school-age population in high school had increased during this period from about one in fourteen to one in nine, and educators were agog at the unprecedented growth and increasing democracy of their schools.

Thus, we see during this period a substantial increase in the number of children and young teens working outside their households, and a more modest number of them going to school. Whole families that would once have worked together on a farm might still be working together in a mill or a mine. The work in the mill might not have been harder than what John Muir had to do for his father, but it was counted—and perceived—differently.

If you were a teenager in Pennsylvania's hard coal region, you would very likely have already been working in the mines for several years. Only a handful of miners' children—mostly daughters—stayed in school for more than the rudiments of reading and arithmetic. (Even writing was viewed as a specialized skill for future clerks and bookkeepers, not for those who expected to go to the mines.) There were laws requiring the youngest children—first those under twelve, and later those under thirteen—to go to school and not to the mines. But as in most states, children were able to begin work once they had grown large enough that they and their parents could lie about their age and not appear ridiculous.

It was unusual for young workers to actually go into the mines to extract coal until they were well into their teens. The mines nevertheless depended on the labor of those in their early teens, or even younger. They opened and shut doors in the mines, took care of mules, delivered messages, and most important of all, worked as "breaker boys."

If you were a breaker boy, you would stand over a series of chutes and belts and remove pieces of slate from the crushed coal as it went past. Because slate and hard anthracite coal are difficult to tell apart, you would have to keep your face very close to the just-crushed rock in order to see the difference. You would breathe coal dust from morning to night. No wonder breaker boys with black faces, black arms, black hair, and black clothes became, by the turn of the century, the poster children for advocates of child labor legislation. They were small and appealing, but with faces that looked more like forty than fourteen.

Still, your parents would have had little choice but to send you to work, and in most mining towns, the mines were virtually the only employer. An 1881 state survey of 141 mining families in Scranton, a city with a more diverse economy than most coal-mining centers, found not one father whose income amounted to more than half of the total family income. In a large number of the instances, the income of the head of household was less than a quarter of that of the entire family. There was one case where an anthracite miner with five children contributed only $80 of the $800 the family made that year.

It would be interesting to know the circumstances of that final, extreme case. Moral reformers often argued that many working-class parents loafed at home while sending their children off to support the family. Although alcohol was most often blamed, some argued that work in the mines and factories was so arduous and soul-crushing that adults were simply worn out. They may have started working as children themselves, and they were no longer able to brave the toil and tedium of dawn-to-dusk labor.

"In an incredibly large proportion of cases," an 1895 study of youth employment found, "the fathers of young wage-earning children not only do not support the family, but are themselves supported by it, being superannuated early in the forties by the exhaustion characteristic of the garment trades, or the rheumatism of the ditcher and the sewer-digger, and various other sorts of out-door workers; or by that loss of limb which is regarded as a regular risk in the building-trades or among railroad hands." The authors added, however, that another cause of child labor, especially among Italians, Germans, and Bohemians, was what they termed "sheer excess in thrift" of parents eager to own property.

In some industries, new machines made it possible for factories to lay off skilled workers and replace them with young men in their teens whose wages were considerably less. As a teenager, you might endanger your father's job. It would be worse, though, if your father lost his job to someone else's son.

Thomas O'Donnell, an immigrant from England who worked in a Fall River, Massachusetts, mill, told an 1883 congressional committee hearing that the heavy machinery he had operated had been supplanted by machines boys could operate. "There are so many men in the city to work," he told the committee, "and whoever has a boy can have work, and whoever has no boy stands no chance."

According to O'Donnell's account, fathers depended on the sons not simply to supplement their income, but to ensure that there would be any income at all. That encouraged families to pull children out of school as soon as they could get away with it, in order to secure a place for the father and other family members at the mill. Fall River had a high school by 1883, but it's not surprising that few children of mill hands were enrolled.

For many immigrants, economic dependence on their young was only the beginning. "The child stands between the new life and its strange customs," wrote Mary McDowell, a social worker who established a settlement house near Chicago's stockyards in 1894. "He is the interpreter; he is often the first breadwinner; he becomes the authority in the family. The parents are displaced because they are helpless, and must trust the children."

It is easier for young people to master a new language than for their parents, and that's where immigrant parents' dependence on their children began. Culture, however, often proved to be an even more important issue, especially among those in their teens. Young people were quick to figure out how to act American, which was something the parents depended on.

But when mothers and fathers saw their children coming of age not as Sicilians, Polish Jews, Bohemians, or Irishmen, it could be heartbreaking. Growing up American often implied a rejection of parents' values, beliefs, and rituals—if not of the parents themselves. Parents and their children were living a paradox. Parents bewailed the ignorance and thoughtlessness of their offspring, but they needed the knowledge they were acquiring about how to survive in the new land. "Now that you have become Americanized," one writer quoted his Italian mother screaming at him in exasperation, "you understand everything and I understand nothing."

McDowell lamented that the religious and social ideals the parents brought with them were quickly supplanted by their children's more superficial ones, shaped, she said, by teachers, politicians, and saloonkeepers. " 'Shut up talking about Bohemia,' said a boy to his mother who was shedding homesick tears as she spoke of the beauties of her old home. 'We are going to live in America, not in Bohemia.' She had a vision of beauty, while she was living in the sordid ugliness of the stockyards. . . ." But that same ugliness was her son's home, the place where he would make his life.

In the biggest late-nineteenth-century cities, many families needed the wages their children could earn, but whole families working for a single employer was rare. Cities like New York, Boston, Philadelphia, and Chicago, which were ports of entry or magnets for particular immigrant groups, offered employers a large number of able-bodied, willing, mature, low-paid workers. The productivity of the factories was high enough and the wages were low enough to displace nearly all children and most teenagers from production jobs. There were, to be sure, many cases of families doing piecework in their homes, creating tenement-based sweatshops. In New York, for example, some Italian families set up virtual production lines to make artificial flowers, with five- and six-year-olds performing the simplest tasks at the beginning, their older brothers and sisters executing finer tasks, and their mother finishing the job.

Nevertheless, by mid-century, more than half of the eleven- to fifteen-year-olds in the biggest cities were neither in jobs nor in school. If you were a big-city youth who needed to make some money, you could work for yourself—on the streets. You might be a newspaper carrier or a bootblack or a messenger, working long, hard days, hustling for a penny or a nickel at a time. At the time, these occupations, known as "the street trades," were thought disreputable, yet people found it very convenient that someone was available to carry a parcel or deliver a message at nearly any hour. In retrospect, the young people in the street trades played an important part of the development of the modern city, with its rich concentration of goods and information to be moved around.

If you wanted to take a step up from the street trades and get a regular income indoors, you could become a cash boy or girl at a big department store. John Wanamaker, Macy's, Marshall Field, and other great merchants were building multistory stores that covered entire city blocks. The stores wanted it to be convenient for buyers to purchase products, but it was risky to allow clerks throughout the sprawling store to handle money before the introduction of the recording cash register. When the clerks had finished writing up an order, they would yell, "Cash!" and one of the uniformed teenagers would run to pick up a basket containing the buyer's money, a sales slip, and often the item as well. The item was brought to a wrapping desk, the money to a cashier, who would place change in the basket, and all would be returned to the customer.

The cash boys and girls were a control and management device that, in a sense, made the big department stores possible. There were a lot of them. During the 1870s, one in three Macy's employees were cash boys or girls. Department stores were the largest single employers of big-city twelve- to sixteen-year-olds. But the cash girls and boys were a short-lived army. In 1902 Macy's installed eighteen miles of brass pneumatic tubes and rendered the cash boys and girls obsolete.

Outside the store, the swarms of young people who made their living carrying messages were decimated by the telephone. They never entirely disappeared, of course. Bicycle couriers—most of them in their twenties—became very visible on city streets during the 1980s, both because of their colorful skin-tight outfits and their death-defying riding style. There will be a niche for such couriers as long as there are cities, and young people are going to fill it. Still, now that information passes instantaneously through wires and the air, there will never again be a place for the armies of young people who were, at the end of the nineteenth century, occupied with carrying bits of writing from one place to another.

"I was only fourteen years of age when I obtained a job in the Marshall Field store in Chicago," recalled Eddie Guerin, a notorious bank robber, in his memoir. "Running about with money appealed to me well enough, though I could not earn sufficient to satisfy myself. Seven or eight months I lasted in the store, and then I became a Western Union telegraph boy. . . . But here again the pay did not come up to my requirements. Night after night I would be playing pool and gambling in dollars, until I began to look for ways and means of raising money. The only opportunity I could see of getting any was to steal. Delivering telegrams as I was to all sorts of private homes and business establishments, I found there were plenty of chances to pilfer small articles. . . ."

Guerin, who at fourteen was living on his own after his mother evicted him because of his vicious habits and unruly ways, was nevertheless successful at obtaining employment during the 1870s. Even in that decade, perhaps the high point of teen messenger employment, the job didn't pay very much. Horatio Alger wrote a novel, *The Cash Boy*, about a young man struggling to support himself and his orphaned sister. The only way he managed to make ends meet was by turning out to be the long-lost heir of a millionaire, an option not open to Guerin or to most others.

Moreover, Guerin had a full range of grown-up temptations open to him. "Chicago of the 'seventies was a wild and woolly affair," he wrote. "Public houses and pool rooms, gambling establishments of all kinds, red-light houses, kept by foreign women, open all night long. . . ." Laws to keep young people from partaking of these pleasures were few and rarely enforced.

Guerin was admittedly not a typical young man. (He later became world-famous for accomplishing the unheard-of feat of escaping from the Devil's Island prison.) His story nevertheless reminds us that the nascent information industries that moved through the streets of the nineteenth-century city intersected with sex and gambling, much as they do on the Internet today. A young man like Guerin might deliver something—and take something else away. The young newsboy or messenger was sometimes lionized as an entrepreneur in the making, at other times demonized as an incipient criminal. Guerin himself noted that he was a cash boy at Field's at around the same time as Gordon Selfridge, the future London department store magnate.

The teens in the street trades were an important clientele for musical shows and other sorts of cheap amusements that arose in the cities at that time. Their patronage of these burlesques offered proof that the young newsboys weren't all working to support a tubercular widowed mother at home, as legend had it. Often in Alger's novels, the hero takes his first step toward reform when he stops going to shows and puts the price of admission in a bank instead.

There was, indeed, a lively working-class culture, much patronized by the young, of cheap theaters, dance halls, amusement gardens, and semiprofessional sports. Working-class youths didn't have a lot of money, but much to the consternation of the upper-class philanthropists who sought to reform them, they spent enough of it on amusements to have an impact. The amount of entertainment available seems almost unbelievable today. For example, in the vicinity of Eighth and Race streets in Philadelphia in 1900, there were more than 100 theaters—none of them respectable.

Indeed, the ways in which young people amuse themselves has been a source of worry at least since the eighteenth century, when elders worried that their daughters were being ruined by too many novels. The emergence of a febrile, big-city, working-class popular culture, patronized largely by the young and reflecting their tastes and desires, galvanized opposition. For the first time, the shows young people liked, the music they played, and the dances they did came to be seen as a threat to the proper order of society. This nascent youth culture provoked a powerful political backlash.

During the 1870s, the moral reformer Anthony Comstock, leader of the New York Society for the Suppression of Vice, succeeded in getting Congress to pass an antiobscenity law later emulated by several states. He argued that low-class urban amusements were the incubators of juvenile delinquency and led to costly public expenditure: "If gambling salons, concert dives, lottery and policy shops, pool-rooms, low theaters, and rumholes are allowed to be left open; if obscene books and pictures, foul papers and criminal stories for the young, then must State Prisons, penitentiaries, workhouses, jails, reformatories etc. be erected and supported." Censorship, he argued, was not merely right but economical.

Comstock's movement, which made little distinction between popular entertainment and criminal activity, focused on threats to the young. The legislation it sponsored, however, prevented everybody—not just young people—from reading banned publications, and it sought to close, not police, the theaters. Although courts and state legislatures eventually gutted most of the legislation passed under Comstock's influence, the fear of working-class culture and its impact on youth has never really disappeared.

As late as 1905, an official of the New York Public Library removed George Bernard Shaw's work from circulation. "Supposing [*Man and Superman*] fell into the hands of a little East Sider," he asked. "Do you think it would do him any good to read that the criminal before the bar of justice is no more of a criminal than the Magistrate trying him? Do you think that would lower the statistics of juvenile delinquency?" Shaw, via *The New York Times*, responded, "Comstockery is the world's standing joke," and attributed the librarian's action to "the secret and intense resolve of the petty domesticity of the world to tolerate no criticism and suffer no invasion."

The "petty domesticity" Shaw denounced was, in fact, the conventional establishment solution to the problems of lower-class urban youth. The bourgeois home seemed the model for reforming working-class children, a belief that placed reformers at odds with parents, priests, and others who resisted efforts to make their children into thrifty, temperate Protestant homebodies.

Some so-called child-savers, such as Charles Loring Brace, became deeply concerned with the lives of the young men found in legitimate and semilegitimate street-based businesses, and occasionally juvenile delinquency became a matter of public concern. Bigger cities led to larger reform schools, though not yet to a separate justice system for the young. Most of the time, however, big-city citizens viewed the highly visible young people toiling and hustling on the streets as a convenience or, at worst, a nuisance. Only at the turn of the century, when technology began to eliminate youths' jobs, did concerns about the welfare, education, delinquency, and labor of people in their teens begin to emerge.

During the late nineteenth century, children and young people in their teens did grueling grown-up work, often in hazardous industries. Did this have the paradoxical effect of making young men and women more childlike—or at least smaller and slower to mature?

Muir believed that the hard work he had done on his father's two Wisconsin farms had actually stunted his growth, and turned him, the firstborn, into the runt of the family. Like many Americans who are middle-aged, I remember parental warnings that some behavior would stunt my growth. Even as a child, I recognized, I think, that this was a phrase that had come echoing down through generations of parents, a living cultural antique. The acts most likely to stunt one's growth, according to parents, were smoking and masturbation. In fact, tobacco is a toxin that can retard the growth of a fetus exposed to it in the womb, and of young adolescents who smoke a lot. Masturbation, on the other hand, seems to have no relation to the growth of the body or any of its parts. "Hard work," another very old saying goes, "never killed anyone." We have seen, however, that it does retard the growth and maturation of pubescent female gymnasts. But did it make Muir and others like him smaller?

The answer to that question isn't clear. What is true is that, during the latter part of the nineteenth century, something was making Americans smaller. The average height of American men born during the 1880s was four centimeters—more than an inch and a half—shorter than those born five decades before. This was true even though virtually every economic measure had increased enormously, even on a per capita basis. This appears to be the only period in the last 350 years when the average American actually became smaller. Increasing disparities between rich and poor produced a small population of the tall and prosperous and a large population of the small and poor. Average heights of Amherst and Yale students were shooting up, even as those of army recruits were going down.

Some of the reasons for this phenomenon seem obvious in retrospect, though they were largely unrecognized at the time. While young people were, for example, protected from going into mines because of the risk of collapses and poisonous gases, many young people suffered from routine, prolonged exposure to toxic substances. The danger of chronic contact with coal dust, tobacco, cotton dust, and other industrial substances became apparent only gradually, and employers often denied the connection.

There were other factors as well. Families that had moved from farms to cities were probably eating less meat. More working women meant less breastfeeding and lower infant nutrition. Higher population densities increased susceptibility to epidemics.

By the turn of the century, the smallest generation of young people in memory reached its teens. During the first years of the new century, the photographer Lewis Wickes Hine began his memorable series of polemical photographs of young workers. Their bodies were tiny, their faces prematurely aged. And at this same moment, psychologists and educators were developing a new view of the human life span in which adolescents were seen not as young adults but as distinct creatures. Having produced a generation of small-bodied young, it was easy for Americans to accept that their teenage offspring were not fully grown.

Still, not every nineteenth-century teenager was getting a start on black lung in the mines or suffering stunted growth from labor in a cigarette factory. Many memoirs of the southwestern frontier—from Kansas south to Texas and westward to California—are romances of the frontier. Their stories are exceptional for their period. If you were in your teens in the West, your family would most likely depend on you. Still, there was plenty of space in which to improvise your life and learn how to be yourself, rather than just another worker in a factory or mine. The stories describe youth as a time of great independence and mobility; these young people loved their horses better than any teenager could love his car.

"I rode like an Indian and at the age of sixteen did not lack for suitors," Mrs. Jack Miles of San Angelo, Texas, boasted to a WPA interviewer in 1941 as she recalled her youth on the range, six decades before. "I led them all on a wild race from one end of the Concho County to the other."

She recounted her teen years as a sort of idyll on horseback, enlivened by dungareed Lochinvars. "The most treasured gift that love could buy for me was a horse, so one of my suitors presented me with a fine steed, which I called Ball Stockings. My young heart swayed mightily toward the donor. Then another suitor presented me with a still finer steed to add to my mount, and asked me to return Ball Stockings to his former owner, which I refused to do. One night a bunch of us were riding out to a dance at the Doak Ranch and I was proudly mounted upon Ball Stockings, riding beside my latest suitor, when *crack!* went a whip across my horse's hips. The jealous donor of Ball Stockings had only meant to interrupt our conversation but he succeeded beyond his intentions, for my adorable Ball Stockings broke into a startled and furious run and I had a race as wildly exciting as my heart could crave, while my anxious comrades flew after me. This settled my interest in the jealous suitor."

Soon afterward, still a teenager, she wed the young man she deemed the most attractive of her suitors. They literally rode off into the sunset, into a marriage where both husband and wife were cowpokes.

Narratives like Mrs. Miles's look back to medieval romance and chivalry, but they also look forward to the California dream of a self-sufficient teen culture of surfboards and little deuce coupes, great bodies and rock and roll. While they don't paint a picture of parents abjectly dependent on their children, they also describe youth as a time of considerable responsibility, sometimes punctuated by wild revels.

"Until I was sixteen years old, I punched cattle about three months every spring and helped with the roundups, branding the calves," recalled Charles L. Weibush, another WPA informant, who was born in Texas in 1873 to parents who had immigrated from Germany. "When we had the roundups, the cowboys would ride around the herd at night and sing the cowboy songs, sometimes just croon them to keep the herd quiet. The thoughts of the beautiful moonlight nights, the big herds and the cowboys, the cool fresh air as the morning dawned gives me a homesick longing for the days of the range. . . . I remember once when we went into a saloon and the boys took their drink, as I was so young they gave me soda water. . . . Soon some cowboys from another ranch came in and had their drinks and one of them saw me and insisted that I have a drink of whiskey with him. When I refused, he took out his gun and began shooting at my feet and had me to dance to keep out of the way of the bullets. To this I did my best and pretty soon my crowd took it up and before they decided the question whether I was to dance or not to dance there was a free-for-all fight with no injury excepting a few swollen eyes and bumps from each other's fists. After they had settled this to their satisfaction, they forgot me and called it a day."

Weibush's nostalgia for the range aside, this seems like the sort of incident that was more fun to tell about later than to actually live through. Still, it's the sort of story that shows up in a lot of memories of the western frontier, fulfilling at least in people's recollections, the function of an initiation. The young person has faced a danger, crossed a threshold, and at a fairly young age, left childhood behind.

This southwestern frontier, with its youthful and disproportionately male population, offered ample opportunities for young people to come face-to-face with their own mortality. "In those days," recalled Miguel Antonio Otero of his boyhood and teen years in Kansas, Colorado, and New Mexico, "everyone, man and boy, carried some sort of weapon at all times. It made no difference what the occasion—even a wedding or funeral was no exception—the weapon, in all likelihood a pistol, would be carried in the hip pocket, or in a holster suspended from the belt." Otero, who became territorial governor of New Mexico, was a privileged young man; his father was a powerful merchant who kept moving his family westward along with the Atchison, Topeka, and Santa Fe Railroad, with which he had a long business relationship. Young Otero's parents kept sending him east to school, but he had a talent for becoming mortally ill—until he could be sent back to the frontier. Otero's lengthy memoir of his frontier youth describes limitless buffalo herds, hostile and friendly Indians, notorious outlaws, wild nights with dance hall girls—a full and eventful life for a teenager working as a clerk in his father's store.

One sometimes wonders whether Otero and other memoirists are remembering movies they've seen, rather than their own lives. Still, some darker aspects come through. It's striking, for example, how many of Otero's contemporaries met violent ends, whether in gunfights, Indian combat, or hunting accidents. He noted offhandedly at one point that suicide was also common.

And although Otero didn't mention it, the atmosphere of multicultural machismo he described must have been very difficult for women, especially young ones. He boasted that he had begun frequenting dance halls from the time he was thirteen or fourteen. Although he didn't describe them as brothels, he is fairly specific about the location of the girls' bedrooms in relation to the dance floor and bar. (They weren't upstairs, as they are in most movies, but along a corridor right off the dance floor. Two-story buildings were not common in the often short-lived boomtowns created by mines and railroads.) He says that fellows his age were expected to go to such places, and that marriageable girls would forgive them for it.

Not very much is known about the women who provided entertainment on the frontier, except that they were predominantly in their teens, and they probably came mostly from large cities, where there was less work for young women than in smaller mill towns. Many might well have become prostitutes in any case; the growth of large anonymous cities in which most people were strangers to one another had made that a growth industry.

Most people were aware at the time that there was a shortage of women in the West. It's not unlikely that some of the girls who went west as dance hall girls had hopes of finding a husband—maybe even a wealthy one—and some did. Still, most men on the frontier wanted women not as a civilizing force but as a sexual outlet. One woman who went to Oregon to teach school recalled the scene as men lined up at the station when the train came in to see women get off. They were visibly disappointed when they saw her, she recalled, because she was no longer in her teens. She felt they wished that she had stayed home. But perhaps young women who got almost no attention back where they came from took pleasure in being the center of attention at the dance hall. While it's not known how many dance hall girls parlayed that role into a more respectable one as wife and mother, Westerners did tend to have standards of respectability that were far more flexible and forgiving than those in force back east.

Indeed, the standards of conduct for young women who were daughters of ranchers and merchants permitted behavior that would have been scandalous elsewhere. The dances they attended were chaperoned. But they were also likely to go on until dawn and end with breakfast. More than once, the chaperones dozed.

Women on the frontier were both precious and useless. They were precious because they were rare, and essential for the creation of permanent settlements. Aside from a few assertive exceptions like Mrs. Miles, they were useless because they weren't thought fit to do the rough, dangerous work that needed to be done. Moreover, many men didn't know their wives very well when they married them, not uncommonly on a rushed trip back home for the purpose of finding a suitable mate.

All of these helped make western families less stable than those of other regions, a distinction that has persisted for more than a century.

Thus, in the towns of a region thought short of wives, there was, nevertheless, a noticeable population of single mothers. Some were widowed, but many were separated or were abandoned by their husbands. There wasn't much work for the mothers to do. But especially in mining areas, their sons were greatly in demand. As in the coal towns, they most often worked outside the mines, tending mules, clearing sluice ways, and performing assorted odd jobs. Unlike their Pennsylvania counterparts, Rocky Mountain teens sometimes staked mining claims and started businesses of their own. They lived in a place through which a lot of money was passing. Through shrewd business dealings, or skill and luck at the gambling table, a fifteen-year-old could become someone to reckon with.

Young men in their early teens also met with success in other fields. One thirteen-year-old Montanan was so successful at selling books that a publisher paid him $1,000 a year to be its representative. Others were merchants or wholesalers. Another ran a ferry, which was an essential, and sometimes lucrative, business.

"I have never seen any children," wrote an Englishwoman traveling in the West in 1873, "only debased imitations of men and women, cankered by greed and selfishness and asserting and gaining complete independence of their parents at ten years old." She was exaggerating, most likely, though she wasn't alone in her sentiments. Stories of ten- and twelve-year-olds sidling into the saloon for a morning snort were commonplace. And don't forget that young man from New York's Lower East Side who went west as a boy, killed his first man at twelve, and died as a legend—Billy the Kid—at the age of twenty-two.

There were significantly fewer children and young teenagers, as a percentage of the population, in the West than elsewhere in the country. They seemed, however, to frighten visitors more. The reason wasn't that they were pathetic, or even so much that they were prone to drink, gamble, and fornicate, though this was a concern. Their most threatening aspect was that, amid a shortage of labor, in places of rapid development and sudden wealth, a boy in his teens could behave exactly the same as a much older man. He might be a successful businessman, a thief, or a drunkard, but he was not a child. He was a full member of his society.

There was, however, as Otero repeatedly discovered, the danger of being sent east to school. Young people who were living grown-up lives on the frontier would immediately revert to a state of immaturity. Sometimes the young person didn't even have to travel that far. As a thirteen-year-old in Alder Gulch (later Virginia City), Montana, Raleigh Wilkerson worked in the mines, but when his family moved to Helena a few years later, he entered school. "From being almost a man," he remembered later "[I] became once more a boy."

In the twentieth century, going to school would become the key teenage experience. Nevertheless, some qualities of these young people of the frontier—their love of mobility, their belief that they were entitled to have a good time, their sudden shifts from dependence to maturity and back—became part of that creature we know as a teenager.

EIGHT
The Invention of High School

When I ran away from home, I never considered that I was running away from school, too. School is very important to me, because I know I need it to reach my dream of working in the film industry.

—RUSSELL LASH in *YO! (Youth Outlook)*, (1994)

Russell Lash was a sixteen-year-old high school student in San Francisco when he wrote a story for *YO! (Youth Outlook)*, revealing that he lived in a car parked in his high school parking lot. His teachers and classmates didn't know he was staying alive by panhandling change on street corners. He dreaded cold nights and feared the exposure of sleeping in public. His accounts of late nights and early mornings echo those found in *Ragged Dick* and other nineteenth-century romances of youths redeemed from poverty. Lash apparently wasn't alone; one of his female classmates slept beneath the front steps of the school building.

In June 1996 another modern-day Horatio Alger tale made the front pages. Camara Barrett, a young immigrant from Jamaica who was banished from his Brooklyn home by his stepfather and an increasingly violent mother, lived first on the New York subways and then in a homeless shelter. What made him newsworthy was that such all-too-common horrors had led not to a life of crime, but to graduation as valedictorian of his high school and a scholarship to Cornell University.

Now, as in Alger's day, stories of young people overcoming terrible odds help those who are comfortable escape guilt about the routine brutality of contemporary life. The problems that gave rise to the nineteenth-century dangerous classes—poverty, substance abuse, poor living conditions, and what they called "disorganized" and we call "dysfunctional" families—are still with us.

Alger's heroes had only to add some good habits—thrift, soap, and Sunday school—to the enterprise they already displayed in abundance. Their modern-day counterparts face an additional obstacle: They have to succeed in high school.

Well into the twentieth century, some young people—those with clear vision, strong ambition, and little patience—were able to bypass high school and move directly into making a living. Now young people who wish to succeed in any sort of legitimate enterprise must make it through high school. To reject your family is one thing; there may be very good reasons. To reject high school is to reject the society as a whole.

High school is the threshold through which every young American must pass. Its classes impart knowledge we believe young people need to become good adults. Its athletics and extracurricular activities provide the principal stage for young people to explore their talents and find their strengths. It brings young people together, providing a fertile ground for the development of youth culture. By enrolling both young men and women, the high school gave teenagers control over their own social life, something that parents controlled before everyone went to high school. Without high school, there are no teenagers.

American high schools are often criticized, but rarely challenged. We have become so accustomed to the idea that high school should be the universal experience of our youth that we don't even consider other possibilities.

This widespread acceptance of the idea of high school was very, very slow in coming. The high school movement did not hit like a tidal wave, but rather like a glacier, slowly insinuating itself into American life. The first public high school opened in Boston in 1821, but New York City didn't open one until more than seventy years later. Only during the 1930s was a majority of what we now term high-school-age youth enrolled, mostly because the Depression had made jobs unavailable. The idea of high school for all took hold only after World War II, a bit more than fifty years ago.

For more than a century, Americans resisted public high schools, sometimes on ideological or political grounds, but most often passively, simply by not making it part of their own or their children's lives. They couldn't sacrifice the young people's income, and they couldn't imagine ways in which it would help them have better lives.

During much of the nineteenth century, there were probably more teenagers working in mines than attending high school. There were certainly many more teenagers working in factories. Through most of the century, there were a dozen on farms for each one in high school. Americans believed in schooling for their children, but they didn't think they could afford to forgo the income that their teenage offspring could provide. If they could not send their own children to high school, they certainly didn't want to pay taxes to support the schooling of youths from families that were wealthier than they.

For many nineteenth-century parents, high school seemed to make more sense for their daughters than for their sons. This was a major reversal of the ancient practice of reserving learning for men. One important reason for this change was that young women had a clear career path from finishing high school to teaching elementary school. Besides, in the sheltering, middle-class home, mothers were expected to serve as their families' most important teachers. Still, the overriding reason for parents to send their daughters to high school was that families did not depend nearly as much on the daughters' income as that of their sons. When, for a variety of reasons, parents had nothing more productive to do with their sons either, high school became, at long last, the normal habitat of Americans in their teens.

High school enrolled only a tiny minority of nineteenth-century youths. Still, the institution's eventual triumph makes its early history significant. By looking at how high schools started, what they might have become, and how they evolved into what they became today, we can unearth the roots of problems that afflict today's high schools.

The United States in the early nineteenth century was one of the most literate places on earth. To the Protestant precept that one's ability to read the Bible was a necessary condition for a pious life was added the republican notion that literacy was essential for participation in political life. In settled places, the common school was a well-established—if haphazard—institution where reading and the rudiments of arithmetic were taught. In frontier places, women did fulfill their role as first teachers to their families. For those who became learned on the frontier—Abraham Lincoln is a familiar example—schooling was a minor part of their lives, but reading was essential. Only slaves, who were usually forbidden to learn how to read or write, lagged far behind.

There was also a tremendous proliferation of institutions called colleges, many of which survive today. College still concentrated on preparing students to become clergymen, and most towns now boasted churches associated with many different sects. This fight for market share among the Baptists, Methodists, Presbyterians, and Congregationalists encouraged the increased number of colleges.

The settlers of Massachusetts had founded Harvard soon after their arrival, and other New England towns almost immediately founded Latin grammar schools to enable their sons to get in. Because these grammar schools were, in many cases, supported by public funds, advocates of public secondary education often cited these as precursors to the high school. Although Boston Latin School survives as part of that city's public school system, the provision of publicly supported secondary schools did not spread beyond New England, and they were not widespread even there.

Most state constitutions trumpeted the importance of education and granted localities the power to make provision for it, especially for the poor. Beyond the basics, though, schooling was a family responsibility. In the South, well-to-do parents engaged private tutors to prepare their children for further education. In the old northeastern states, some private academies began to specialize in preparing young men for college—often for specific colleges that counted on them to supply students. Farther west, newly created institutions such as the University of Michigan created preparatory departments and branches. In some cases, these colleges themselves became, in effect, secondary schools. Indeed, the distinction between secondary and higher education was not very clear.

It seems irrational, in retrospect, that there should have been such a growth in colleges while secondary schooling remained rather primitive. Such a conclusion assumes, however, that education is a system that runs from basic literacy to higher learning, as Horace Mann advocated successfully in Massachusetts.

Such a view of education was slow to win acceptance in other states. Some attempts to set up public schools provoked court challenges arguing that the use of public funds for universal schooling was unconstitutional. Education advocates won in every state where there was a challenge. At the time, however, the outcome of the battle was far from obvious. The creation of an educational system raises fundamental questions about how society benefits from schooling beyond the basics, what that should be, and how it should be achieved.

Most people viewed education—and particularly preparation for college—as a family responsibility. Even the common school was viewed more as a cooperative project than a truly public one. In most states, the only area in which government took a responsibility was in schooling the poor. This tended, by most accounts, to be grudging. As early as 1817, the chairman of Philadelphia's school committee criticized charity schooling as "not only injurious to the rising generation, but benevolent fraud upon the public bounty." Such criticisms—that "it's bad for the recipients, and besides it costs us too much"—have been applied to many welfare programs since then.

In an effort to save money, Philadelphia, like other cities, adopted an approach introduced in England a few years before by the reformer Joseph Lancaster. This was, in essence, an application of industrial principles to the school. A typical Lancastrian school enrolled students from five to fifteen years of age, organized on a "monitorial" system in which the older students bore responsibility for teaching and disciplining the younger children. Most of what was taught was memorized and recited in unison throughout the day. "By the aid of monitors," Lancaster declared, "one master can teach one thousand boys." The system was also notable for completely barring corporal punishment, the chief disciplinary tool in nearly all schools of the time. Public humiliation was employed instead.

Lower labor costs were the chief attraction. Philadelphia decided to go ahead on the basis of a report stating that 3,000 poor children could be educated at a public cost of $3,000 a year. In 1818 Lancaster actually moved to Philadelphia, where he founded a model school intended to train the (very few) teachers that would be needed in Lancastrian schools in the city and elsewhere. Lancaster tended to move on before the failure of his methods was fully appreciated. By 1823 he was off to South America. Still, the Lancastrian schools established in Philadelphia, New York and elsewhere helped establish a precedent for public schools that enrolled students in their teens.

In 1828 the labor movement, which was to become one of the strongest advocates of public education, spoke out for the first time. Leaders of newly formed labor unions organized a labor party in Pennsylvania, and the creation of a public education system was one of the chief planks of its platform. It argued that public education was not merely a matter of charity, but of right. This particular labor movement did not survive, but many of the unions that arose later in the century took up the cause. Labor leaders realized that the employment of youths and children kept wages down, and pitted fathers and sons in competition for the same jobs. More idealistically, the 1828 labor platform declared that a system of public instruction, "calculated to impart equality as well as mental culture," was the only way that the aim of "republican liberty," declared in 1776, could be realized.

This was not a new argument in 1828, and it has been repeated countless times since. It is fundamentally a political argument. If schooling is not universally available and accessible, then America is tacitly accepting the division of society into classes and castes and turning its back on its own fundamental principles. Moreover, there were places in Europe, particularly in Germany, that were doing a far better job of making sophisticated schooling available to a much larger portion of the population than was the United States.

The academies, which were diverse, responsive to student demand, and flexible in their programs, embodied the free market aspect of American ideology. At a time when industrialization was beginning to increase economic inequality, at least some people believed that a different kind of school was needed to maintain social cohesion.

Once you accept this premise, however, difficult questions arise, some of which have never been answered satisfactorily. Is democratic schooling different in its nature from the traditional approaches? How do you reconcile a belief in equality with the realities that students have vastly different aptitudes and that no school will be able to develop each of them equally? How much schooling is necessary to maintain equality, and when does it become either a subsidy for the well-to-do or an imposition on students' time?

These questions are easiest to resolve on the primary level, where it is possible to establish a broad consensus on what children should be taught. It is far more difficult to define what a democracy owes its young people—and has a right to demand from them—once they reach their teens.

Even in the early years, high schools had a symbolic importance. Most were housed in large, well-appointed buildings at the center of town and staffed by well-paid male teachers, rather than the poorly paid women who taught primary school. High schools were both a point of pride and an inviting political target.

★   ★   ★

The Boston English Classical School, soon to be known as Boston English High School, opened in 1821 and was the first of its kind. Its goal was not to prepare students for Harvard, but rather, as was said at the time and ever since, to "prepare them for life." Preparing for Harvard meant, almost exclusively, reading and translating Latin texts and mastering Latin grammar and learning algebra and geometry. The declared aim of Boston English was "to give a child an education that shall fit him for an active life, and shall serve as a foundation for eminence in his profession, whether Mercantile or Mechanic."

Preparing for life is a good deal less predictable than preparing for college, and it presents a challenge to schoolmaster and student alike. It meant mastering what educators then called "English" or "new" subjects. By this they meant just about everything—not only English language, rhetoric, and literature, but chemistry, physics, astronomy, geography, history, civics, and possibly bookkeeping, manual training, and pedagogy, too. They were still new subjects because, although Harvard, whose president was a chemist, had loosened up a good deal on its Latin requirements, Yale had not, and neither had some other influential colleges.

Boston English wasn't for everyone. There was a difficult entry examination, and during the student's three years, he had to study a wide array of subjects. These included what we would call academic courses, such as literary criticism, logic, moral and political philosophy, and ancient and modern history, along with a few applied courses, such as navigation and surveying. The course of studies bore a strong relationship to that advocated by Benjamin Franklin for his academy eight decades before. A young man coming out of the high school at the age of fifteen or sixteen was assumed to need no further schooling and to be well prepared to play a leading role in the community.

In the years that followed, high schools began to proliferate throughout the northeastern United States. Boston started an English high school for girls in 1825, then quickly closed it down. Soon other cities and towns either started such schools for girls, or more commonly, admitted both boys and girls to the same school. There was a good deal of public soul-searching about the moral consequences of mixing the sexes at such a dangerous age. Except in the largest cities, where there were enough students to justify investing in separate high schools for each sex, economies of scale triumphed over scruples. The sturdy virtue of young American womanhood was relied upon to keep matters from getting out of hand.

Many of these early high schools were little more than a room in an elementary school where older students who hadn't found a job, or who thought they needed to learn some geometry or German, were accommodated. Others, though, were extremely ambitious attempts to create a new kind of school equal to the college in stature, but better attuned to the needs of the society America was becoming.

We have become accustomed to seeing the high school diploma as the lowliest of credentials, and to judging high schools by their success in fitting their students for higher education. For nineteenth-century teenagers and their families, high school was a substantial sacrifice of time and earning capacity. High schools could not satisfy students, their parents, or their political constituencies simply by keeping young people in custody. They had to produce.

In 1849 James J. McElhone, sixteen, a newly minted graduate of Philadelphia's Central High School, moved to Washington, where he joined several other graduates of the school as members of the official corps of congressional reporters. During the next two years, as Congress struggled with the issue of whether slavery should be allowed to expand westward, McElhone took down what the lawmakers said word for word in Pitman shorthand he had learned at the high school. He and his fellow teenage reporters thus created the official record of some of the most celebrated congressional debates in history.

Central taught the skill—then known as phonography—along with such other practical subjects as navigation and bookkeeping and less immediately applicable ones like Latin, Spanish, Anglo-Saxon, trigonometry, and history. The school, which opened in 1838, was modeled on the United States Military Academy at West Point, and it boasted a distinguished faculty. The celebrated painter Rembrandt Peale, for example, was its first teacher of art—and of handwriting. Like many other early high schools, it was sold to the citizens of the city as a replacement for college. Indeed, only a handful of colleges at the time were so rigorous in their offerings, particularly in the sciences. This all-male school was called "the people's college," but their graduates were expected to have a more practical bent. Within only a few years of its founding, completing the course at Central had become a valuable employment credential.

The school was attempting a combination that is difficult for us to imagine today, at least in part because we tend to take teenagers' intellectual capacities somewhat less seriously than did Central's teachers. They believed they could produce educated men who were highly employable, all by the age of sixteen.

Oddly, the achievements of James McElhone and his fellow alumni were a source of early controversy, mostly because such practical subjects as shorthand were not considered appropriate for a serious school. In 1854 there was a movement within the school to drop the subject, followed by a public countermovement to preserve it. The school's principal, a bookish man whose own pet project was the course in Anglo-Saxon, defended the course. He said some of the graduates, "not yet turned of twenty, are making more money by phonography and reporting than the principal of the High School, after having given himself for more than twenty years to his profession." It was, perhaps, an odd moment to angle for a raise, but what comes through more strongly is a rueful ambivalence about the scholarly and practical goals of his institution.

Shortly after, when the antiimmigrant Know-Nothing Party took control of the Philadelphia city government, Central was forced to stop the teaching of all foreign languages—even Anglo-Saxon. Unlike the private academies and the older colleges, high schools were vulnerable: They were publicly funded, novel, expensive, and elitist. What high schools teach and how they are run are still matters of frequent local controversy. Most of the time, though, the attention is short-lived and the public's interest moves on to other issues. That's what happened at Central in the 1850s. Most of the faculty was rehired, and within a few years, all the foreign languages except Anglo-Saxon were back in the curriculum.

A different, more benign, form of local interest in the high school was particularly evident in smaller cities at examination times. Most schools taught primarily through recitation, and they tested their students orally. Examinations afforded the schools a chance to show off to the public. Citizens could observe the examinations and, in some places, ask the students questions of their own. These public examinations served more as exhibitions staged by the schools to prove their efficacy than real examinations. Later, student oratorical contests, plays, concerts, and above all, sports also evolved from the desire of schools to show off and that of parents to see their children perform. Rarely, though, do current high school events concentrate, as those of early high schools did, on the content of what the pupils studied.

Graduates of some of the better big-city high schools went immediately into highly responsible work. Indeed, it was commonly assumed that high school graduates could deal with the real business of the world better than those who had graduated from college. High schools were urban and businesslike, while colleges and academies were pastoral and detached from the real work of either farm or city.

Thus, it was possible to have a career like that of Elihu Thomson, an 1870 graduate of Central. While still a student, he had done experiments with electrical dynamos and had begun a scientific society at the school, which produced a number of discoveries published in scientific journals. Upon graduation at seventeen, he became a member of the school's faculty and continued his own research and publishing. He left the school ten years later to help found the General Electric Company. (Thomson, the French electronics giant, is named for him.)

Unlike today's technologically precocious teens, Thomson apparently didn't have to put up with being called a geek or a nerd. If he had been, though, it probably wouldn't have mattered much, any more than it would to the similarly self-directed young people still found in high schools. Thomson was clearly someone who would have succeeded in any era. As a high school graduate, he had more education than many of the other technological entrepreneurs of his era. Indeed, Central was, at the time, an outstanding institution that commanded greater respect and helped students establish more important contacts than most colleges then or now.

Contemporary teenagers, whether in high school universities or community colleges, must learn how to live within a highly bureaucratized society. During the 1980s, when anthropologist Michael Moffatt studied Rutgers University students, they told him repeatedly that they viewed the school's size, impersonality and frequent administrative lapses as useful training for adult life. The awareness that they were only numbers on a computer—a computer that's usually down—provided a common experience and made the students skilled at adapting to, and occasionally subverting, the system.

The small world of which Thomson and his classmates were part seems very attractive in comparison. Yet, it's important to realize that it was artificially small, the result of a sorting and examination process in which most young people did not participate. During its first century of existence, the American high school was an undemocratic institution: Children of members of the professional and managerial classes were overrepresented, while those of skilled artisans attended in proportion to the general population, and those of factory workers, farmers, miners, and most immigrant groups and racial minorities were either severely underrepresented or completely missing. The schooling may have been free, and the young people may truly have desired it, but many families were not free to give up the income of their young during the high school years.

Booker T. Washington described how, as a child recently freed from slavery by the Emancipation Proclamation, he looked forward to going to one of the schools established, mostly by Northerners, throughout the South. These schools enrolled all ages, opening opportunities for literacy that were closed to nearly all slaves until then. Despite Washington's desire, his family needed money, and his father had little choice but to send him to the mines. (Later, when he turned sixteen, Washington left his family and walked to the Hampton Institute, which let him earn his tuition by being a janitor.)

Similar stories were repeated among urban immigrants and native-born whites as well. Political opponents of public high schools, though characterized by educators as yawping barbarian hordes, often had justice on their side. The poor, they argued, had no hope of being able to afford to send their children to high school. They needed their older children's wages to support the household. Yet, they paid for the high schools through their taxes. High school was what would now be called an upper-class entitlement, subsidized by those who could not afford to take advantage of it. This argument prevailed only rarely, usually during brief periods when populists or insurgents managed to gain control of city councils or school boards. While the beneficiaries of high school tended to be small in number, they belonged to the groups that tradition-ally dominate local politics.

Educators, for their part, tended to assert that they were part of a movement that was growing every year. In one sense, they were correct. Absolute numbers of high school students kept growing, new schools opened, and new teachers were hired. That did not mean, however, that high schools were truly becoming more democratic. There were decades when the percentage of Americans in their teens enrolled in high school actually dropped. Immigration was adding millions to the U.S. population, but most high school educators did not think of immigrants' children as potential students, or even as Americans. Thus, they asserted the benefits of universal secondary education, even as they were actually losing ground.

And just what were those benefits? What would students—and the society as a whole—derive from universal secondary education? Then, as now, educators and even politicians rarely addressed the question directly, regarding the answer as so self-evident it needn't even be discussed. In fact, the answers were not, and are not, obvious. Schoolmen might proclaim in speeches and editorials the ability of public education to realize democracy by preventing caste. Parents, however, usually see this issue differently. They hope that their children's schooling will at least preserve their position within the class system or, even better, enhance it. Parents aren't all that interested in democracy. They're looking not for equality but for an advantage.

The high school movement rested on the idea that serious schooling could also be practical. Yet, parents usually demand that their children's public education be similar to that received by the privileged in private schools.

The conservatism of middle-class parents helps account for a paradox. Education is constantly buffeted by new theories, approaches, methods, and systems. Teachers and administrators adopt or revile these novelties, they become the subject of books and conferences, and the public hears that schooling will be revolutionized (or destroyed) by the latest pedagogical fad. Yet, in the classroom, change comes slowly, if at all.

By the mid-nineteenth century, elite secondary education was centuries behind the times. Schools that prepared young men for the top universities taught little but mathematics and Latin grammar and literature. The most important recent innovation was an increased emphasis on Greek. As some reform-minded schoolmen liked to point out, Latin was, in medieval times, a vocationally oriented subject. It was the language of the church, of law, of medicine, and of learning in general. Anyone who wanted to enter the clergy or the learned professions could expect to be using it for the rest of his life.

But why should someone living in a mostly Protestant, English-speaking country a hemisphere removed from the classical world have to spend years learning the grammar of the ancient Romans and not a minute on that of his own language?

Educators had an answer: mental discipline. The point of schooling, they maintained, was not to learn subject matter. The specific things people need to know varies, they said, from person to person and from time to time and can be learned after one has picked a profession. The proper role of secondary schooling, the argument went, to discipline the mind to adapt to any kind of study—in other words, to learn how to learn.

This is not a ridiculous argument. Even by the late nineteenth century, it was obvious that knowledge was exploding, while once precious skills were becoming worthless. Too narrow a course of study was, and is, a dangerous temptation. The trouble was that it wasn't at all clear that studying Latin really disciplined the mind. In fact, psychological research was beginning to suggest that there were no fundamental disciplinary subjects. Different people have different aptitudes for learning different things, and students who did poorly in a Latin and mathematics curriculum might excel at other things.

Latin kept bouncing back because parents demanded it—even if their children had no expectations of going on to the colleges that required it. In America, democracy translates into the dream that anybody can be a member of the upper class. Everyone involved may have believed that Latin study was the key to mental discipline, but it survived as a status marker.

When Boston founded its English High School, it offered no Latin whatsoever. The Boston Latin School had been teaching that for more than 180 years. The new high school was to be something altogether different. A few other cities, especially those in New England, did the same, at least for a time. Philadelphia's Central—with chemistry, Caesar, and shorthand maintaining a contentious coexistence in the curriculum— was more typical. Its point was not that it was "English," but that it was public, the capstone and chief ornament of a city-supported school system. As with coeducation, economy played a part in pushing toward a comprehensive school. It would be very expensive to maintain two schools—one practical and terminal and the other for the college-bound. If one of these schools were to be sacrificed, it would logically be the Latin school. Private academies existed to do the job, and the parents of college-bound students were presumably able to pay. Nevertheless, the chief constituency for the high school was, as the populist critics pointed out, the professional and managerial classes, who wanted high-status careers for their children. The response, then as now, was generally to provide different courses of study within each high school, to meet different goals and provide for different students.

For example, students entering Providence High School in Providence, Rhode Island, during the 1850s and 1860s studied in one of three distinct departments. Boys could choose between the classical course and the English and scientific course, while girls were confined to the girls' department.

Despite the frequently expressed fear that education could break a woman's health and shatter her nerves, the course for girls was quite rigorous. When a young woman arrived for the first of her four years, she studied botany, rhetoric, Latin, algebra, physical geography, and general history. In subsequent years, she added French, the history of English literature, natural philosophy—i.e., astronomy and physics—physiology, chemistry, geometry, English prosody, intellectual philosophy, geology, and moral science. By the time she had finished, she had taken three years of mathematics, four of Latin, three of French, and six of science. And young women were far more likely to finish than their brothers.

That's why the boys entering the English and scientific department were overloaded with subjects, the most useful of which were clustered in the first two years. A freshman at Providence studied algebra; the history of Greece, Rome, France, and England; physical geography; English grammar; declamation; composition; vocal music; physiology; and bookkeeping. It's difficult to assess how seriously these subjects were taught. Early high schools tended to offer too many courses for their teaching staffs to handle, thus engendering the dependence on textbooks that continues to this day.

These two curricula have their quirks. Why, one wonders, four years of singing for boys and none for girls? Still, they were more or less typical of high school offerings elsewhere, and they express an intention to acquaint students with a wide range of serious, demanding studies.

What's shocking today is the course pursued by the boys in the more prestigious classical course. They didn't sing, nor did they study any science, any modern languages, any English literature, grammar, or composition, or any mathematics other than algebra. After their first year, everything they studied was Latin or Greek. Providence was at the heart of one of the most rapidly industrializing areas in the country, but the most prestigious and popular course at its high school was still proudly medieval.

★   ★   ★

The 1860s mark a dividing line in American education. Before the Civil War, students could treat primary schools and academies as if they were wells: They could attend them when they wanted to get the schooling they felt they needed. After the war, high schools supplanted academies and primary schools were divided into grades. Students met the school's schedule. The student became more like a piece of timber, which, if one followed a suitable design and carried out a series of operations in the proper sequence, could be crafted into a serviceable table or chair. Public education was beginning to be conceived as a system, and as a proper role of government.

The reasons for these changes had little to do with the war. The major educational impact of the Union victory was that slaves were, for the first time, able to go to school. Unfortunately, the South's impoverishment made it an educational backwater. Schools attended by both whites and blacks lagged behind the rest of the nation for more than a century.

Still, one of the most important innovations in American education was enacted by Congress in 1862, when the war was going badly for the North. This was the Morrill Act, creating the so-called land grant colleges. Under the law, the federal government pledged to contribute grants of land to the states to set up colleges dedicated to the improvement of agriculture. The law did not directly affect secondary schools, but it helped change the way people thought about education, especially in the Great Lakes and Plains states.

Until then, higher education was mostly for those who wanted to be clergymen or doctors or lawyers. Professors had little to teach about what to do when your soil's fertility ran out, or how to cope with pests or get more corn per acre. Even when academies and colleges were in rural settings, they stood as arcadian refuges from the backbreaking life on the farms that surrounded them.

The Morrill Act must rank as one of the most imaginative pieces of legislation the U.S. Congress had ever produced. Research on agriculture was virtually nil. Because rich agricultural land seemed unlimited, there had been little incentive to find ways to make the land more productive. By establishing agricultural research as a national priority, the Morrill Act changed that mentality and encouraged the creation of new fields of study.

Less directly, but even more radically, it made a link between advanced knowledge and farm life that hadn't existed before. Schooling was previously thought beneficial primarily to those working in cities. At the time, American cities were experiencing a period of explosive growth. Yet, the vast majority of Americans were still on the farm.

While students in Cambridge and New Haven were assiduously delving into the antique, a new kind of knowledge was emerging in East Lansing, Ames, Columbus, and Madison. Because much of this new learning concerned dairy cattle, pigs, and alfalfa, it was easy to condescend to these new "cow colleges." Yet, students could integrate the science they were learning in college with their years of experience on the farm, and work at the cutting edge.

Still, both students and colleges faced an enormous obstacle. The students were ill-prepared academically, and there were few secondary schools in farming regions. As a result, most of the colleges opened preparatory departments to help students make the transition between their haphazard primary schooling and the demands of the college. In some states, the land grant colleges were active in helping cities and towns set up public high schools with curricula developed by professors at the college. Unlike earlier high schools in Boston, Philadelphia, and other northeastern cities, these high schools were intended to prepare for college, not to replace it.

While many of the land grant institutions grew during the twentieth century into huge and highly prestigious institutions, they remained small during the nineteenth century. Few young people from farms went to high school, much less college. They were still too valuable to spare. Nevertheless, the Morrill Act established a principle that publicly supported schooling could benefit everybody, even those who expected to spend their lives on the land.

This principle, admirable as it is, came at a substantial price: It established the colleges as the arbiters of what high school ought to be. The high school movement began at Boston English with the goal of preparing for life, not necessarily for college. But, as we have seen, parents frequently pushed public high schools into another direction—that of preparing students for the elite East Coast colleges. Now professors at the land grant universities were literally designing the secondary schools of the Midwest to serve as feeders for their institutions. The colleges were, admittedly, of a new and useful type. Still, this conception of high school as a preparatory institution is a surrender of the original ideal: that a high school education should be immediately useful both for the student and the community at large.

The role of universities in the shaping of American high school education has, most often, been informal and advisory. Over the last 125 years, they have, nevertheless, had a greater influence than such other powerful constituencies as business, labor, or even high school teachers and administrators. That's because the quality of a high school is usually judged by its graduates' admission into and performance in college. Preparing students for life is an admirable goal, but you have to wait a lifetime to see whether you have done a good job. College admission, by contrast, is an immediate indicator of success.

Yet, there has always been a problem with such a college-centered approach. In the nineteenth century, the majority of high school students didn't even graduate. Today the great majority do, but it's still not clear that pointing everyone toward college is the best way to help each student develop his or her particular talents and prepare to make a living.

One attempted solution was the establishment of manual training high schools in a few large cities. They did not teach any specific trade, but instead attempted to provide a basic familiarity with a number of handcraft and industrial skills so that students could discover their interests. As has frequently been the case in such efforts, ties between school and the workplace were weak and what the students learned was often decades out-of-date. Labor organizations opposed manual training because they feared that the students would be semiskilled strikebreakers, trained at public expense. Besides, many manual jobs were already open to teenagers. There was a far greater incentive for a father to find his son a job that would help pay the rent than to sacrifice the income by sending him to a school offering inferior training.

Another approach to the problem of what to do with the non-academic student was the junior high school. The long-forgotten promise of the junior high school was that it would teach the basic verbal and mathematical skills most students would need in everyday life by the age of thirteen or fourteen. The setting of the junior high, which resembled high school rather than elementary school, and the presence of more male teachers made it seem a more mature experience, one that might help the student feel comfortable working in an office.

Unlike manual training schools, junior high schools were very popular with the labor unions and the working class generally. They offered the hope of upward mobility without forcing the student to give up too much time. By the 1880s, it was clear that white-collar jobs provided greater opportunity for a lifetime of income than did laboring jobs. In jobs that depend on physical strength, the peak earning years are during one's twenties, while white-collar jobs are rewarding for a far longer period. Among white-collar jobs, the greatest demand was for those that involved fast, accurate arithmetic skills. The growth of industry had created an explosion of clerical tasks and recordkeeping. The typewriter and new calculating machines were beginning to supplant old-fashioned handwritten ledgers. While such skills were little taught in high schools, the hope was that junior high schools could provide at least a foundation for surviving in this climate.

This vision for the junior high school was to prove short-lived, however. As educators fine-tuned the high school as a preparation for college, so did the junior high school turn into little more than a preparation for high school.

Yet another view of the role of the high school devalued its educational mission and proposed instead that it should be an instrument of social control. In a society being transformed by industrialization and immigration, the school seemed to some the only hope for preserving traditional American values.

"I regard the high school, next to the church, as the chief barrier against communistic and socialistic inroads from the howling sea of an ignorant and unprincipled civilization." So wrote an anonymous editorialist in the *Pennsylvania School Journal* in 1879, in words that sound oddly familiar to anyone who experienced a Cold War education. This rhetoric, nearly four decades before the Russian Revolution, is a good example of a recurrent strain of thought in American education. Even if schools teach nothing useful, the argument goes, they still keep Americanism alive, and thus keep the peace. Edward A. Ross, a Stanford professor, wrote in 1901 that the school is "an economical system of police," succeeding religion as "the method of indirect social restraint." He urged that educators keep this true role of the school to themselves. "To betray the secrets of ascendancy is to forearm the individual in his struggle with society," he wrote.

One way in which the schools fulfilled their mission of social control was to become increasingly involved in the nonscholastic aspects of their students' lives. The 1890s brought an explosion of extracurricular activities—athletics, dances, student newspapers, and "school spirit."

In 1900 a student correspondent in the Somerville, Massachusetts, High School *Radiator* did a comparison of papers throughout the country: " 'What a vast difference,' some will say, 'must there be in those papers from such far distant states.' How different school life much be in Oregon and Texas from that in prosaic Massachusetts. But no, it is not so. Everywhere among high school students the same feeling exists; the American spirit of freedom, of good fellowship, and of patriotism. Everywhere is the same attention and interest given to athletic sports."

At the turn of the twentieth century, high school was becoming, as the student paper noted, a standardized experience. The Oregon students were taking the same courses at the same time as their Massachusetts counterparts. This was no accident. It was largely the handiwork of the Committee of Ten, a prestigious panel whose 1894 report on curriculum shaped the high school experience decisively and for a long time.

The committee's chairman and guiding force was Harvard's longtime president Charles R. Eliot. Before taking on the challenge of rethinking the high school curriculum, he had called for new thinking about how to make schooling more efficient. Public education wasted altogether too much of its students' time, he argued. He once had two people read aloud all the books read during six years of schooling; it took forty-six hours. All the math problems could be completed in fifteen. Students could do more younger, he concluded, and keep up the pace through high school.

Such views might well have led Eliot toward a radical reform of secondary schools, or at least made him sympathetic to the junior high school advocates' aim of practical, expeditious education. This did not happen.

Rather than examine high schools on their own terms, Eliot and his nine colleagues looked for ways to increase the depth and breadth of college preparation in high school. They convened conferences on each of the high school subject areas to get recommendations on how much should be taught, and when. The committee and its affiliates did not consult with the businessmen who would be their students' likely employers, or with rural educators. There was only token participation by public high school teachers or administrators. Instead, the Committee of Ten functioned essentially as mediators among scholars and administrators at colleges, universities, and a few elite secondary schools. The group scarcely addressed the fact that few high school students went on to college or even finished high school. One member did, however, assert that preparation for college is the same as preparation for life.

The collective demands of the college professors heard by the committee would have taken far more than four years of students' time. The committee made some trims here, some major cuts there, and came out with a compromise—a model four-year curriculum. It specified the courses students would take during each of their four years and the time devoted to them. The committee presented it as merely an example, not necessarily a standard. Yet, because it had the imprimatur of several of the heads of several prestigious universities, high school principals accepted it as a recipe for success. Most regional differences disappeared. The committee's report defined the modern high school in considerable detail. And although similar bodies revisited the issues later, the Committee of Ten's imprint endured. During the 1960s, my own high school course schedule was very similar to that promulgated by the Committee of Ten seven decades before.

The Committee of Ten's report was a step in the creation of an educational system that is, in some respects, a wonder of the world. Throughout the twentieth century, high school would become, as many earlier educators had hoped, a universal experience. But by enshrining the universities as standards setter for the high schools, the committee ignored the needs of the majority of high school students without academic ambitions.

The high school movement began in a spirit of democratic pragmatism, with the goal of providing a new kind of education to help all kinds of students adapt to a changing world. But that has only rarely been attempted. High school remains, as it always has been, the weak link in our educational system because Americans have never been able to agree on what it should accomplish. The principal reason high schools now enroll nearly all teenagers is that we can't imagine what else they might do.

NINE
Dangerous Adolescence

> Never has youth been exposed to such dangers of both perversion and arrest as in our own land and day. Increasing urban life with its temptations, prematurities, sedentary occupations and passive stimuli just when an active, objective life is most needed, early emancipation and a lessening sense of both duty and discipline, the haste to know and do all befitting man's estate before its time, the mad rush for sudden wealth and the reckless fashions set by its gilded youth. . . .
>
> —G. S. HALL (1904)

Granville Stanley Hall, psychologist and college president, didn't invent the American teenager. But his vision of adolescence as a beautiful and perilous time still exerts a powerful influence over the way we see the young. As the founder of the study of adolescent psychology, he pioneered scientific inquiry into nearly every facet of the youthful mind and body. He also created persistent, destructive clichés.

Over and over throughout his two-volume, 1,400-page tome, *Adolescence: Its Psychology and Its Relations to Physiology, Anthropology, Sociology, Sex, Crime, Religion and Education*, Hall deplores what he called "our urbanized hothouse life, that tends to ripen everything before its time." Like many other members of the professional and middle classes, Hall worried that young people were precocious in all the wrong ways. They grew up too quickly, expecting to make too many of their own choices, getting old before they had fully dealt with being young.

This was a concern that emerged with great force around the turn of the century. It animated educators, lawyers and judges, labor unionists, politicians, and a substantial concerned citizenry. It helped produce laws calling for longer compulsory attendance in school and for distinctive legal procedures to be applied to young people. It helped push the relatively elite high schools of the nineteenth century to become more inclusive. Over time, it contributed to the decline in child and teenage labor and the passage of laws to regulate it.

Such major changes in the way Americans have lived their lives were spurred by forces in the economy, technology, and politics, both in the United States and elsewhere. They cannot be attributed to one book, even one as imposing and ambitious as Hall's. Nevertheless, the book played a catalytic role, giving various groups an intellectual basis for their campaigns to get young people off the streets, out of the factories, and into the schools.

It also helped create a distinct field of intellectual inquiry. And it offered an image of the adolescent as a figure at once beautiful and troubled, passionate and feckless. Hall's envy and fear of the young come through on nearly every page, along with his empathy, an honest effort to understand, and declarations of sexual attraction he may not have known he was making. This volatile mix of ideas and emotions was present at the creation of the modern adolescent, and it haunts us all, young and old, still.

To today's reader, *Adolescence* seems to say more about Hall's own psychological quirks than about the issues of adolescents. In page after page of run-on sentences, he expounds on the radiance and joy of youth, and how easily it can go disastrously astray. In an effort to give his work a sound scientific basis, he reports on countless studies from each of the fields named in his title. He often seems to jump to conclusions that go beyond what the data suggest. As Hall himself noted, many of the studies cited did not deal specifically with adolescents. That's not surprising because, before Hall, adolescents were not generally seen as a discrete group worthy of their own studies. Subsequent studies of this age group disproved much of what Hall asserted as fact, but these might not have been undertaken if Hall had not first defined the field.

It's hard to imagine that many read Hall's book, even when it was new. It was destined, and perhaps intended, to have its impact secondhand, as its ideas filtered down to educators, clergy, youth workers, lawmakers, and parents. Still, it would not be accurate to conclude that the popularization of Hall's ideas stripped them of those elements that we find ridiculous or offensive today. On the contrary, Hall's ideas managed to overcome his writing style and academic apparatus precisely because they responded to the fears, conditions, and prejudices of his time.

For Hall, the growth of the adolescent was not simply physical, or even mental and emotional—important as these may be. Hall expounded a kind of mystical Darwinism with a racist tinge. Each person must recapitulate human evolution, he argued, and some individuals, and even cultures, face the danger of being arrested at a stage lower than that which the best of humanity has reached. He argued that children eight to twelve years old had reached "the age of maturity in some remote, perhaps pigmoid, stage of human evolution, when in a warm climate the young of our species once shifted for themselves independently of further parental aid."

During most of the nineteenth century, fourteen-year-olds were viewed as inexperienced adults—people with energy who could work a full day, but might, if not supervised, fall into vicious habits. For Hall, the fourteen-year-old was no longer a child, but still very far from being an adult. The physical development of the young person—the increased muscle mass and widened shoulders of men and the pelvic development of women—was not, he said, evidence of maturity. These adult physical features were, he said, armaments in a struggle to achieve a higher state of being. "Youth awakes to a new world," Hall wrote, "and understands neither it nor himself."

He presents such a litany of risks in adolescence—ranging from pre-occupation with one's skin to suicide—that readers wonder how they survived its perils. Hall saw the adolescent quest as so powerful and so urgent that normal young people are likely to show symptoms that would, in adults, be evidence of severe mental disorder. (This is still widely believed, but untrue.) At the same time, Hall often overdramatized the difficulty of adjustments in adolescence that most people make quite painlessly. The reader who has had a relatively sane second decade feels lacking, perhaps less fully evolved.

Like many of his contemporaries, Hall used a Darwinian framework to justify racial and ethnic prejudices. He argued that the stage of evolution that had produced the adolescent struggle for maturity was quite recent, and somehow not yet working as effectively as the mechanism that produced the pygmy and the eight- to twelve-year-old. This ascent had been led, he assumed, by those who did not live in tropical climates, implicitly Europeans. Obviously, this had implications within American

society. For example, he argued that the contemporary city and its temptations had made the petty crime of young people a matter for broad
social concern, rather than a matter an individual family could control.
He compared this to the freeing of the slaves in 1863, after which, he
said, petty thievery that had previously been tolerated and controlled by
slaveowners was suddenly cause for punishment. Hall's clear implication
was that these people of African descent had probably been better off as
slaves. Likewise, he argued that adolescents should be under strict parental
control and should not be allowed to spend the money they earn.

In the context of his time, though, Hall was almost a liberal. He
thought it beyond question that "the negro is excitable and lives in his
emotions," but he admired the efforts of former slaves and their children
to educate their young. He believed in an industrial education for black
youths, as he did for lower-class whites, and worried vaguely that the
blacks were doing a better job of it in the South.

Still, Hall's book succeeded because it spoke to its time. Hall provided a reason for his readers to feel superior to immigrants, Negroes,
and adolescents—and cause to be afraid of them.

Concern about adolescents, and the crime and social problems associated with them, is perennially focused on other people's children—especially those with a different native language or skin color. It's a recurring
theme of American life that the generations are ethnically different from
one another, and that the rising generation appears less "American" than
the one doing the perceiving. At the turn of the twentieth century when
Hall was writing, new immigrants of southern and eastern European
descent and black migrants from the South were changing American
cities. Hall observed, in a worried tone, that "the senescence of the
original American stock is already seen in abandoned farms and the infe-
cundity of graduates." He noted that America's population increase was
entirely due to immigrants and their children.

At the turn of the twenty-first century, young people of Asian and
Latin American descent make up a larger percentage of the current teenage generation. At both times, the birthrates of the new Americans were
higher than those of the natives. When a disproportionate number of
the young look different from the population as a whole, they present a
special threat.

Hall wrote about adolescence as a stage in human life, not about
American adolescents. Still, his book coincided with and supported a

movement toward much more aggressive assimilation of Americans of all ethnicities and classes into a middle-class pattern of life.

School was seen as virtually the only safe place for the menacing young. Only a few years after Hall's book appeared, a prominent Missouri educator called for all young people to be "sentenced" to high school, both to give them essential training and to protect the public. He argued further that universal schooling would provide an opportunity for weeding out the feeble-minded, who could be sent to institutions that would ensure that they would not have children of their own.

Hall's work appeared at a moment when America was self-consciously trying to civilize itself. Chicago's World Columbian Exhibition of 1893 set off a vogue for grand classical civic centers and parkways— vast formal spaces amid the smoke and congestion of burgeoning industrial cities.

Suburbs, meanwhile, were spreading along railroad lines. Factory owners retreated to Cotswold cottages, worshipped at Episcopal churches named for eccentrically English saints, and started cricket clubs. Likewise, Ivy League universities erected Gothic buildings contrived to look ancient the day they opened. All was part of an effort to assert the existence of an ancient elite that deserved to dominate.

We have seen how, half a century before, some relatively prosperous families understood that, in a fast-changing society, schooling would help their children to adapt. At the time Hall was writing, those who spent their second decade studying rather than working were still a small minority. Still, they were the third generation to have had this experience, and they tended to view this life course as somehow "natural" or even essential.

For the working class, however, childhood's end was occurring sooner and sooner. While upper-class children—mostly native-born with English antecedents—were marrying later and having fewer children, the working-class white immigrant and black populations were following the traditional path of early marriage and many children.

While Hall claimed to be examining a universal phenomenon, he wrote entirely from the perspective of the professional class of his time. By doing so, he reveals that concerns about adolescents and their proper development often resolve themselves into concerns about class. Hall seems at times to be willfully oblivious to what was going on around him. He writes about the crime of the working class and the sex of the working class, but gives little attention to the work of the working class. Even though large numbers of children, and most young people, were in the workforce at the time he was writing, he says next to nothing about it. His laundry-list subtitle mentions physiology, anthropology, sociology, and education, but one appropriate discipline, economics, is entirely absent.

While Hall had nothing to say about industrialists who worked adolescents for long hours in unhealthy conditions, he condemned feminists, who he said failed to recognize that establishing "normal periodicity" should be the preeminent task of a year or more of girls' adolescence. Hall was less interested in public health than in "secret vice," which he confidently asserted to be increasing. Hall spoke little about the way in which young people were exploited, and still less about the ways in which young people were functioning as fully competent members of their society. He worried instead about what he called arrested development, perversion, and hoodlumism. The society was helping to cause these conditions, he argued, because parents, educators, and clergymen did not understand adolescents.

Hall's implicit position on working youth was that they shouldn't be working because they had too much else to accomplish. Mostly, though, he wasn't interested in the working class. The miracle was that five decades after Hall's book appeared, the United States was wealthy enough so that nearly everyone—even those Hall ignored—could afford to have a troubled adolescence.

One reason why Hall's work was welcomed, especially by educators and youth workers, is that it provided a theory for many of their own efforts. High schools, for example, were increasingly attempting not merely to instruct their students, but also to offer competitive sports, extracurricular activities and dances, and other social events. By making the high school into a self-contained world, they were attempting to counter the allure of outside sports clubs and commercial amusements dominated by the working class. Increasingly, clubs and teams organized by young people were supplanted by organizations run by adults, often operating at a national or even an international scale. The overall effect was to make people in their teens into a distinctive group that was less often relied upon—even by itself—to take responsibility and make its own decisions.

Team sports, for example, began as voluntary, informal organizations.

They occasionally had a loose affiliation with a school, but more frequently allowed participation by the great majority of young men who were not in school. Many teams were fielded by the first of the adult-dominated youth organizations: the YMCA.

By the end of the nineteenth century, however, students, not workers, dominated team sports. What had begun as a recreation after a day of physical labor became a substitute for labor, an outlet for young men who had otherwise forfeited physical activity in favor of mental training.

The schools were spurred to take control of team sports by the growing popularity—and brutality—of football. Male students at Detroit High School, for example, had organized a football team in 1886, with the approval of the school's principal, who praised the players' "manliness" and "gentlemanliness." The team played other high schools, teams from the University of Michigan and other colleges, and YMCA and other nonschool teams. In 1893, however, a rumor surfaced that one of the players on a Toledo YMCA team that defeated Detroit High was a professional wrestler and prizefighter. It's not clear whether the rumor was true; what is important is that people became upset. Parents worried about their sons' safety. School administrators saw a threat to their school's reputation. Students felt aggrieved because it seemed unfair to have to face such a formidable opponent, whether or not any specific regulation was violated.

Many colleges had already moved to take control of their own athletic programs, which were becoming a big business as well as a focus of student life. It seemed that high school clubs, with their younger players, needed at least an equal measure of protection. Moreover, sports in high schools faced heavy public criticism. Already, parents and educators were criticizing "win at all cost" attitudes that, some said, were leading some young men to drink whiskey during the games to kill the pain and allow them to keep playing. The only way sports could be saved would be for educators to take them over, to ensure that they would be preserved, as one Michigan principal put it, as "an excellent avocation for a studious growing boy."

Defenders of high school sports promoted another innovation during the 1890s. They decided to encourage young women to attend as spectators, and later as cheerleaders, in the belief that they would exercise a refining influence. "The presence of the fair sex has without a question a telling effect upon the character and result of every game played," the Somerville, Massachusetts, High School *Radiator* reported. Before the end of the decade, female students at many high schools had created a role for themselves at the games: leading the crowd in cheers. That quintessential teen couple—the football player and his cheerleader girlfriend—was on hand to greet the twentieth century.

During the last decade of the nineteenth century, both high schools and their students adopted many of the institutions and traditions associated with colleges and universities. They launched drama clubs, glee clubs, photography clubs, and student government. Students could buy class rings. They published yearbooks. High schools were, after all, supposed to serve as the people's college, and very few students were likely to go on to higher education, or even to graduate from high school.

Thus, while the senior prom, where it existed at all, was not the elaborate event it would become, hazing freshmen was a widely popular ritual. Usually, Greek letter fraternities and sororities organized the hazing. These had been founded and run by students and were usually tolerated by school administrators. In many schools, the Greek letter groups started the sports teams and drama clubs, and in many more, they organized dances and other social activities. While some teachers and principals voiced objections to the secret and exclusionary nature of fraternities and sororities, they generally took the position that the students were free to organize their own social lives. Besides, the students seemed safer at a fraternity party than at some public amusement spot.

After 1900, however, this laissez-faire attitude faded. Increasingly, educators believed their role was not solely, or even primarily, to impart knowledge. The high school, they said, served the community by taking substantial responsibility for every aspect of a student's life, in and out of the classroom. William McAndrew, the principal of Washington Irving High School for Girls in New York City, said in 1910 that his students wanted "to study the social amenities that make life more pleasant and friendship more enjoyable," and added that doing so would help curb crime, divorce, and immoral literature and entertainment.

Many schoolmen said that what the school ought to do is develop character, but what they meant by character was itself undergoing a rapid transformation. It was less a matter of personal integrity than of being a good team player. The purpose of schooling, said Newark, New Jersey, superintendent Addison Poland in 1913, is "not individuality but social unity . . . unity which results in efficiency and is rarely, or never, obtained except by and through uniformity of some kind." He added, "Children must be taught to live and work together cooperatively; to submit their individual wills to the will of the majority; and to conform to social requirements whether they approve of them or not." This was known as the doctrine of social efficiency, an idea that gained such currency it was written into several state constitutions. Although it is no longer a cause, it played an important role in the creation of the all-encompassing, full service high school we know today.

For proponents of social efficiency, fraternities were not merely petty and cruel, but a threat to the high school itself. Some even called them un-American. For them, the students were no longer individuals, exercising the freedom to associate guaranteed in the Bill of Rights. They were, instead, adolescents whose successful negotiation of a dangerous stage of life would determine the nation's future.

The difficulty in eradicating fraternities was that young people in their early teens appeared to have a great need to be part of a group. The principals counted on that to help inculcate "school spirit," an intangible substance first identified during the 1890s, which they sought to use as a building block of social efficiency. Groups get their identity, however, only through contrast with other groups, which is a large part of the attraction of fraternities and sororities. The battle against these organizations was fought school by school over a period of decades. School administrators eventually win all disputes with students, if only by outlasting them. High school fraternities and sororities were amazingly persistent, killed in some places only by the Great Depression and the upheaval it brought to the schools. They have since been revived here and there, informally or in secret, but they have never regained the importance they once had.

Even as high school won increasing acceptance among the middle class, however, some worried publicly that the experience was "feminizing" the young men who attended them. They spent their days in classes where girls were in the majority. Many of their teachers were women. And except for some salutary rough-and-tumble on the playing field, they were sitting around all day.

"Feminized" high school students were a cause for concern, the worriers argued, because the privileged young men who were able to attend high school came from what they viewed as America's natural leadership class. The well-known nature writer Ernest Thompson Seton wrote that the typical American youth had once been a farmboy, strong and self-sufficient, yet "respectful to . . . superiors and obedient to . . . parents." He had been succeeded, Seton said, by "flat-chested cigarette smokers with shaky nerves and doubtful vitality." He added that it was particularly important to toughen the native-born, so they wouldn't be bested by immigrants.

Sports were part of the answer. Under the influence of President Theodore Roosevelt, the rough-riding ex-weakling, exercise in school became more systematic and was labeled physical education. However, physical education did not become a state-mandated high school course in most localities until World War I, when many new inductees proved unfit. Exercise was also thought to make men more patriotic. "Did you ever hear an unhealthy, un-American doctrine proceed from a normal, healthy body?" asked E. Dana Caulkins of the National Physical Education Service in a wartime speech to educators.

Seton and others, including Roosevelt, offered a solution to the vitality problem outside the schools. This was an import from England: the Boy Scouts. The Boy Scouts of America was established in 1910, two years after its British counterpart. Unlike Hall, who said people in their teens were not yet adults, though no longer children, scouting explicitly aimed to prolong boyhood. Its chief target was the children of the native-born middle class that some of their leaders worried publicly was committing "race suicide." Their hope was that by creating artificial gangs who learned woodcraft and camped in the outdoors, they would halt the precocity that was making the high schools, one leader argued, rivals to Sodom or Babylon.

One of the organization's publications spoke to the group's middle-class constituency when it warned that a Scout must grow up "free from every blemish and stain—then and only then will he be fully equipped to . . . fight the battles of business life." *The Boy Scout Handbook* was more explicit where stains—and power—came from, when it spoke of the need to conserve "the sex fluid . . . that makes a boy manly, strong and noble." Boys should be manly but not men.

Clergy made up nearly a third of early Scoutmasters, and churches sponsored more than half of all Scout troops. Religion itself was beginning to be seen, like the schools, as a feminine thing. Churches supported the Scouts as an effort to win back the males. That's one reason the leaders of the Boy Scout movement opposed the founding of the Girl Scouts. The suggestion that girls could do more or less the same things as the boys contradicted their goal of creating a uniquely masculine institution.

The Boy Scouts of America has been an enduring institution, but it is one far more closely identified with preteenage boys than with teenagers. The founders were hoping to prolong the boyhoods of older youths, but from the very beginning, scouting was most attractive to the boys of that gang-forming, roughhousing time of life whose duration its organizers were seeking to extend. When a young man no longer sees himself as a boy, he's more likely to reject the organization.

As Hall's concept of adolescence gained general acceptance, more and more states passed laws compelling school attendance during the teen years. To be sure, some of these laws had immense loopholes. In Pennsylvania, for example, youths thirteen to sixteen had to be in school—unless they could produce certificates attesting that they had attained third-grade competency in reading and arithmetic. Even if they had not reached that modest level of learning, the going rate for obtaining such certificates, regardless of competence, was 25 cents. (A quarter, admittedly, went farther in those days.)

Although enrollments still represented a small minority of potential students, educators spoke, sometimes proudly, sometimes fearfully, of their mission to educate "the masses, not the classes." They sometimes convinced themselves that they held a central position in society. Melvil Dewey, the library organizer and member of the New York Board of Regents, declared that the high school principal should be "the educational bishop of the community."

People at the time were less likely to pay attention to a pedagogical prelate than to a captain of industry. Some businessmen still saw schooling of people in their teens not simply as a luxury, but as an experience that would hinder students from being good workers. In 1892 James P. Munroe, a Massachusetts paper manufacturer, said schooling "offers to boys and girls wholly unfit for secondary education a temptation to exchange the actual benefit of remunerative work at fifteen years of age for the doubtful advantage of a training that can have no direct bearing on their life work and which, at the time it occurs, may do decided harm."

Businessmen, nevertheless, became at least inadvertent promoters of secondary schooling when they invested in capital equipment and reorganized their production, distribution, and retailing. Over several decades, such business decisions put a lot of young people out of work. By the time they had run their course, the high school, not the factory, the store, or the street, was the habitat of young Americans. Most educators ignored these transformations of industry and commerce. They wanted to believe that young people were in school because parents and students saw the value of education. They rarely acknowledged that for many of students, and even their parents, high school was a very reluctant second choice.

Sometimes the jobs young people do are in the most innovative areas of the economy. This was true of the Lowell factory girls and Rocky Mountain entrepreneurs. Today personal computers and the Internet have created new jobs in which young people, who have come of age with the technology, have an edge. In 1996 *U.S. News and World Report* ran an article estimating that there were 58,000 teenagers doing such work, along with a comment from an expert deploring the phenomenon. He said they were in danger of burning themselves out in work before they have had proper education. Somehow young people who excel in substantive, well-paying jobs always seem more threatening than those who take the marginal jobs we rely upon teenagers to do. That's because we have been convinced by Hall and his followers that it's dangerous for young people to be precocious.

In most times and places, young people work in jobs demanding low skills, paying low wages, and often, lasting for a short time. This is true of contemporary teenage jobs as supermarket baggers, fast-food employees, retail assistants, and other low-level service jobs. It was also true of young people at the end of the nineteenth century. The chief difference was that, instead of being a reserve labor force for the service industries as they are today, they occupied the least skilled, most precarious niches in manufacturing, shipping, and mining.

We tend to imagine that the Industrial Revolution brought large-scale machine production virtually overnight. Although that's more or less what happened in textile manufacture, many other major industries employed large numbers of workers who used rudimentary machines to do repetitive tasks for low wages. These were the industries that employed many workers in their teens and younger. There was, moreover, a threefold increase in child labor throughout the South during the 1890s, when many industries opened their first plants there.

The campaign against child labor was one of the great crusades of the time, and one sign of Hall's influence was that exploitation of fourteen-year-olds came to seem as shocking as that of ten-year-olds. A moral boundary had been moved. People stayed younger longer. Hall's conception of adolescence helped shape the arguments against child labor. It was not, however, what ended young people's employment in the factories, warehouses, mills, and mines.

Increasingly, such innovations as an efficient nationwide railroad network and new techniques of packaging and advertising were creating an advantage for uniform products manufactured and marketed on a regional or national scale. It no longer made sense, for instance, to make bottles in factories where teams of two men and two boys essentially made all the bottles by hand. This system was supplanted by semiautomatic machinery, displacing one of the boys from a dangerous job of carrying molten glass, then the second boy, and then one of the men. By the time of full automation, one man would oversee machines producing many times as the output of the earlier four-person teams.

Industry after industry was installing machinery that performed the workers' tasks more predictably. A few decades earlier, when some employers would hire an experienced man only if he had a few low-paid, unskilled children to offer, by the early 1900s, employers wanted fewer workers with higher skills. These industrial skills were not taught in high school. Employers increasingly recognized that the training they gave workers on the job was an investment that helped them realize the benefits of their new technology. Short-term, low-skilled workers had become bad business.

From about 1900 on, despite the continuing activity in the urban street trades that many at the time saw as a barometer of youths working, the trends in child and teenage employment were downward. By 1917, the Chicago settlement house worker Edith Abbott wrote: "The most convincing argument for the extension of child labor laws is to be found in the fact that at present there is so little demand for the labor of children under sixteen years of age that it is impossible for more than a small percentage of the children who leave school at the age of fourteen or fifteen to find employment."

Because Abbott was a child advocate, her views might be suspect. But the following year a textile industry publication made the case even more baldly. "The labor of children under fourteen years of age is not only inefficient in itself, but tends to lower the efficiency of all departments in which they are employed," the *Textile World Journal* declared. "Also children of fourteen to sixteen years, worked on a short-time basis, are scarcely more efficient and have a disorganizing effect in the departments where they are utilized. Because of these facts and entirely apart from humanitarian consideration, large numbers of southern mills will not re-employ children of these ages."

People in their teens are described by the mill owners, or at least by the writer, as disruptive—just as Hall had argued. Two decades before, southern textile mills had been among the largest employers of children and people in their teens.

Despite the changing labor climate, however, the passage by Congress and state legislatures of comprehensive, loophole-free laws banning the employment of children and those in their early teens took decades of struggle. The first national law was passed in 1916, but it was declared unconstitutional in 1918. A constitutional amendment was introduced in 1924, but it failed to win adoption.

The strong business opposition to child labor legislation seems to contradict the overall trend away from employing the young. It's important to recognize, though, that one of the chief advocates of child legislation was organized labor, which had a clear interest in shrinking the workforce and removing low-wage competition to its membership. Business saw a clear interest in being able to have such a cheaper workforce available to it, even if it chose not to use it.

One argument that opponents of child labor legislation made frequently was that it would force government to interfere in the sacred relationship between parents and their children. Today that argument seems sanctimonious and hypocritical, especially when made by an employer who really wanted only to keep his wage rates low. Nevertheless, it contained a kernel of truth. There were still many parents who depended on the income of their children to make ends meet. Even though the law was said to be for their children's benefit, if it left the family with too little to eat, the child would suffer along with the others.

Only during the Depression, when nearly all work opportunities for younger teens had disappeared, was it possible to win a working-class constituency for child labor laws. By the time they were effective, child labor laws simply recognized changes that the economy and the society had, over nearly four decades, made on their own.

*   *   *

Juvenile delinquency is a subject of constant concern and intermittent panic. During the first years of the twentieth century, the emerging idea of the adolescent had a great impact on how people understood youth crime and how society sought to deal with it.

Hall's fear was that young people were getting old too soon, and that seemed especially true of juvenile offenders. Mere boys appeared to be transgressing in ways that only hardened criminals would have only a generation before. Just as in the 1990s, there were chilling anecdotes to support this thesis. Then, as now, some young people committed shocking crimes that helped convince people that something had to be done. The difference between now and a century ago is that our solution to the problem is to dismantle or weaken the institution our predecessors created.

The creation at the turn of the twentieth century of a juvenile justice system whose jurisdiction included most people in their teens was one of the great causes of the so-called progressive era. For young people, however, it was, at best, a mixed blessing. Their presumed incompetence protected them from a number of dangers, but at the expense of long-established rights.

Still, the juvenile court movement began in idealism and outrage. Benjamin E. Lindsey, a young lawyer recently arrived in Denver, was assigned to defend a couple of burglars. He was led down a long corridor to what he called a cage, where he found two boys playing poker with a safecracker and a horse thief. He didn't say whether the boys were in their teens or younger. But they had been locked up with their hard-bitten criminal companions for more than sixty days. The youths had learned to gamble from the two older men, "upon whom," Lindsey recounted, "they had come to look as great heroes." Lindsey protested to the warden of the jail, who agreed it was "a damned outrage," but he added that because of prison crowding and the boys' parents inability to post bond, he had little choice.

"Here were two boys," Lindsey wrote, "neither of them serious enemies of society, who were about to be convicted of burglary, and have felony records standing against them for the remainder of their lives. And pending the decision of their cases . . . the state was sending them to a school for crime—deliberately teaching them to be horse thieves and safecrackers. It was outrageous and absurd."

Lindsey's handling of the case, which was not so much a defense of the clients as an indictment of the state, won him attention locally. He went on to become the best-known advocate for creating a separate system of courts for defendants sixteen and under.

The juvenile court he founded and presided over in Denver in 1900 was not the nation's first. Chicago had begun one a year earlier. Boston had begun separate trials for juveniles in 1875, and since the 1820s, many cities and states had had houses of refuge, training schools, industrial schools, juvenile farms, truant schools, and ships on which young offenders were jailed separately from adult criminals. Thus, the situation that shocked Lindsey was one that nearly all states had systems to prevent.

Lindsey did not propose merely to punish young people differently, or separately, for the crimes they had committed. He sought, rather, to place young people in an entirely different relationship to the law. The traditional goal of the legal system is to be detached and impartial, and trials seek to determine guilt or innocence and to punish appropriately. Lindsey's vision of a juvenile court was, in many ways, the opposite. It would become part of the lives of the young people who came in contact with it. The rules of evidence would not apply. Lindsey said he simply assumed that just about everyone who came before him had committed the offenses of which they were accused. He further argued that determining guilt or innocence was not the court's goal. Rather, it was to encourage, induce, or coerce young offenders into changing their conduct so that they would not grow into adult criminals.

A key to this approach, Lindsey and many others believed, was secrecy. If the young people behaved responsibly during a long period of probation, it would be as if the earlier transgression had never taken place. Incarceration was an option, of course; it was what made the coercion work.

The proceedings were held without lawyers or juries, under the auspices of the judge, who often had at his disposal probation officers, youth workers, and representatives of child guidance clinics, which had evolved along with the juvenile courts. The legal theory was that the judge was delaying trial in an attempt to settle the problem, more as in a civil court than in a criminal one.

This theory of juvenile justice had wide appeal and great influence. By 1909, ten states, including several of the most urbanized and populous ones, had separate juvenile court systems, and by 1945, all states did.

In return for the possibility of waived punishment and a clean record, the young person gave up most constitutional rights. These included the right to counsel, trial by jury, adherence to the rules of evidence, cross-examination, public view of the trial, and the right of appeal. Only in 1967 did the U.S. Supreme Court find that juveniles had a right to due process of the law, a decision that helped spark the trend toward trying and punishing young people as adults.

There is little doubt that Lindsey and the other early advocates of separate, highly discretionary juvenile justice systems were well intentioned. In their expanded view of adolescent incompetence, they were simply in tune with their times. Like many youth reformers before them, they were quick to take on a paternal role with young offenders. Lindsey even corresponded with and visited them in their neighborhoods. Parents were, when possible, required to be present at the juvenile court proceeding because their cooperation was deemed essential. There was, in addition, a hope the court might also improve the parents' conduct as well.

The juvenile courts tended to exert a form of parental influence that was middle class in tone. Studies suggest that working-class families discipline their children much as the criminal justice system does adults: They determine whether the young person has done the misdeed, exact punishment, and then move on. The juvenile justice system, at its outset, operated on a more emotionally complex, taxing, and time-consuming basis of establishing trust, examining behavior, setting standards for approval, registering disappointment when they are not met, and exacting punishment that's intended to improve.

Harvey H. Baker, the first judge of Boston's juvenile court, compared himself to a doctor. A youth whose offense is minor and routine might see the judge only once, he said, "just as a physician might do in the case of a burn or a bruise. If the offense is serious and likely to be repeated or the conditions surrounding the boy are such that he is likely to have a serious breakdown or if the cause of the difficulty is obscure, he is seen by the judge at frequent intervals, monthly, weekly, or sometimes even daily, as with the patient and the physician in case of tuberculosis or typhoid."

Pioneer judges like Baker and Lindsey seem to have run themselves ragged making so intensely personal a system work. Because this is an approach that depends so deeply on establishing relationships with the offenders, their families, and their probation officers and others working with them, it is not one that can easily be bureaucratized. Still, as cities continued to explode in size, and legislatures and city councils expanded the definition of misbehavior, bureaucracy was inevitable.

Although Lindsey and his fellow reformers saw commitment to an institution as the last resort, it soon became the punishment for even the smallest transgressions. Hanging out on corners or simply behaving in a way that authorities interpreted as "sullen" was enough in some cities to bring young people to the attention of juvenile courts. Often, faced with working-class and immigrant parents whose approach to child rearing was different from their own, judges were quick to find a family to be unfit.

Young women from immigrant families faced a particular problem. Because many worked, they came into contact with men, and they often frequented dance halls and other popular amusements. Equally often, their families saw such conformity to changing American standards as evidence of sexual wantonness. Nearly all the states had vague prohibitions against "willful disobedience," a display of which could result in the young woman being brought before the court. Frequently, it was her parents who had made the complaint, in order to scare her into behaving more traditionally. But on many occasions, the young woman would find herself committed to a female reformatory, and her parents, regretting their action, would be unable to get her freed. The court saw the parents as part of the problem.

Hall and his vision of adolescence galvanized a movement to keep the young from growing up too rapidly. This struggle to extend immaturity has had a mixed legacy. Few today would wish to return to a time when ten-year-olds were in the mills and the mines. Nevertheless, near universal schooling of people in their teens, though widely accepted as desirable, may not always be the best way for young people to use their time.

The juvenile justice system, one of the key achievements of this antiprecocity movement, has been disassembled bit by bit in recent years, and the constituency for saving it is small and embattled. The result is a situation in which crimes that only juveniles commit—such as curfew violation or purchasing cigarettes—are increasing. Meanwhile, protections of the young—such as isolating the young from adults in prison and keeping their court proceedings confidential—are being eliminated.

And the identification of adolescence as a distinct stage of life led the creation of youth culture, which has allowed young people freedoms over social and sexual lives they did not have before. Young people and their concerns dominate popular entertainment, which provides a source of joy and opportunities for expression.

Yet, thinking about the young almost as a separate species segregates adolescents and sets them up as a target for social engineering. Hall loved the young, but his most enduring legacy was to make youth a problem.

TEN
Dancing Daughters

Amory found it rather fascinating to feel that any popular girl he met before eight he might possibly kiss before twelve.

—F. SCOTT FITZGERALD,
This Side of Paradise (1920)

There's something about her, a femininity—no, a sheer femaleness that's going to make trouble.

—"WARNER FABIAN,"
Flaming Youth (1922)

For girls, "Do you spit or do you swallow?" is a typical seventh-grade question.

—DR. CYDELLE BERLIN,
Founder of the Adolescent AIDS prevention program
at Mount Sinai Medical Center, New York,
quoted in *The New York Times* (April 5, 1997)

The debutantes who opened their mouths to a Princetonian's kiss on the eve of World War I seem models of propriety compared with 1997 junior high school girls eager to learn the etiquette of fellatio. At the time, though, the phenomenon of well-bred young women—"twentieth-century women of the luxury class," as the author of the sensationally popular *Flaming Youth* called them—kissing freely and for pleasure was a deeply shocking phenomenon.

Anxieties about young people in their teens frequently revolve around sexuality. And that inevitably means female sexuality. Even in Puritan New England, there were girls who said yes, sometimes with the excuse that Satan had tricked them into doing so. But men did not expect to marry girls who said yes. During the nineteenth century, especially, upper- and middle-class women were expected to be angels of their households, civilizers of men, not their playmates. Lower-class girls were another sort of creature, a necessary, though sometimes frightening, outlet for males' sexual desire.

What was new in the second decade of the twentieth century obviously wasn't female sexuality. Rather, it was that women of all classes were beginning to express it. These tentative beginnings gave way in the next decade, the 1920s, to a youth culture defined by the newly recognized sexuality of girls who used to be "good." Changing female sexuality demands changed male sexuality as well. It also seems to need a new kind of music that drives older folks crazy—in this case, jazz.

They weren't called teenagers yet. The girls were "flappers," a disreputable English word that Americans brought back from World War I and gave a new meaning. There was no comparable term for young men—"sheiks," "smoothies," and "lounge lizards" described some, though not all of them. The whole generation, from high school students to collegians, was known as "youth," a word usually modified by an adjective that was censorious, titillating, or best of all, both at once.

Flaming Youth, a book whose shock value has worn away with the years, provided the most popular phrase of condemnation, but it also suggested that you couldn't take your eyes off modern young women. Written under a pseudonym by Samuel Hopkins Adams, of the presidential Adamses, the book was dedicated, thesauruslike, "To the women of the period set forth, restless, seductive, greedy, discontented, craving sensation, unrestrained, a little morbid, more than a little selfish, intelligent, uneducated, sybaritic, following blind instincts and perverse fancies, slack of mind as she is trim of body, neurotic and vigorous, worshipper of tinsel gods at perfumed altars, fit mate for the hurried, reckless and cynical man of the age, predestined mother of—what manner of being?"

That might seem to sum it up until you consider the movie titles of the era: *Gilded Youth* (1915); *Wild Youth* and *A Youthful Affair* (1918); *Eyes of Youth* (1919); *Blind Youth, The Flame of Youth, Heart of Youth, The Soul of Youth, Youthful Folly* (all 1920); *The Price of Youth, Reckless Youth, Youth Must Have Love, Youth to Youth* (1922); *Daring Youth, Flaming Youth, Madness of Youth, Youthful Cheaters* (1923); *Sporting Youth, Wine of Youth, Youth for Sale* (1924); *Pampered Youth, Passionate Youth, Wings of Youth, Too Much Youth* (1925). That last title might be seen as the cry of a supersaturated market, though the following year did produce both

Fascinating Youth and *Thrilling Youth* before the mania finally started to wane.

In the 1920s we are, for the first time in our story, in a society we can easily recognize as a precursor to our own. Although a majority of young people weren't yet in school, there was a widespread belief that they ought to be, and would be soon. Well-developed, truly national news and entertainment media were enraptured by the styles of young people, who were among their most avid watchers, lookers, and listeners. Young people were, for the first time, setting styles in clothing, hairstyle, music, dancing, and behavior on their own, and both adults and children looked to them as leaders. Young people were organizing their own social lives, and adults felt powerless when they rejected previous standards of propriety as irrelevant to modern life. Certainly, people in their teens had had fun before, but in the twenties it became a right.

"It was the war, you know." That's how characters in countless twenties novels explain a hedonism spawned by a sense of disillusion that is, in most cases, unearned. Certainly, the World War, as it was called, seems a clear dividing line between the ringletted girls and awkwardly lovesick, calflike boys in the novels of Booth Tarkington, and the faster-moving young first chronicled by F. Scott Fitzgerald. Fitzgerald's observation was that only privileged youths were kissing without commitment before the war, while those of the middle class still believed that a kiss indicated a proposal of marriage. "Only in 1920," he wrote, "did the veil finally fall—the Jazz Age was in flower." Not so incidentally, that was the year Fitzgerald's *This Side of Paradise* was published.

It's easy to understand why Fitzgerald and many others saw the change as sudden and spurred by the rich. But lots of other people, ones who hadn't been to Princeton, saw the change coming from the other direction. The girls, they said were dressing like prostitutes and, worse, painting their faces like them. In England, the word "flapper," which originally meant a young duck or partridge, became a slang word for "prostitute" in the late nineteenth century. By World War I, however, the commercial connotation had faded, while the sexual innuendo remained.

The fascinating, horrifying young, so Main Street opinion had it, took their music from the jungle, their style from the street corner, and their morals from the gutter. Today such statements sound like the ranting of fuddy-duddies, but they contain a kernel of truth, perhaps a larger kernel than did Fitzgerald's account. The behavior that won so much attention for privileged women during the 1920s had been pioneered over the previous three decades by women of lower social classes. They were women from the classes from which prostitutes usually came, but they were not prostitutes. They had brought female sexuality out into the open.

Many people, especially men, really seemed to believe that there are two sorts of women—one type well bred and undersexed, the other poor and morally loose. The first group included one's wives and daughters; the other group was very convenient when you wanted to escape from your ideal home.

The middle-class female during the last quarter of the nineteenth century was, rhetorically at least, a frail and sheltered being. Her social life was rigidly monitored and controlled through an etiquette in which carefully chosen, respectable young men were allowed, by other female relatives and the girl herself, to ask if they could call. These meetings were calibrated by length and seriousness, and nearly always chaperoned—though if the courtship seemed headed rapidly toward marriage, the chaperones found other things to do.

This highly sheltered social situation was seriously undercut when, as often happened, she attended high school with males her own age. This created the setting for informal meetings between the sexes that G. S. Hall, among others, deplored. He saw no problem with coeducation in primary schools and colleges, but he believed that changes in female physiology during those years were so dangerous that girls might be kept out of school, "put to grass" as he put it, for a couple of years.

Toward the turn of the century, the same people who worried that schooling was feminizing men were equally worried that it was simply exhausting women. "Many a one has traded her birthright of health and strength and happy and useful living for a mess of pottage made of sheepskin and wrapped in blue ribbon," said the Boston educator Stratton T. Brooks in 1903.

There was, during the late nineteenth century, a general belief that the health and vigor of young women were inversely related to their wealth and beauty. One commonly diagnosed ailment of females in their teens was chlorosis, signaled by a faintly green appearance of the skin and palpitations of the heart. Associated by some with the loss of menstrual blood, one researcher summed it up as "the anemia of good-looking girls." Such girls were described by another researcher as being largely blonde, with pink skin, fine hair and features—a look associated with privileged young women of British descent, and definitely not with most immigrants. The green of the skin was so subtle as to be invisible to many observers, and the disease was largely self-diagnosed. Girls would go to the doctor and complain that their blood must be out of order.

Joan Jacobs Blumberg, who has studied anorexia and bulimia, two of the more recent diseases of adolescent girls, says the chlorotic girls' symptoms were probably real. She says that many well-brought-up girls avoided eating meat because they thought it stimulated sexual appetites, and even led to nymphomania. If the nutrients found in meat weren't replaced, it could lead to a form of anemia. (This doesn't speak to the matter of the young women whose families couldn't afford meat. Perhaps they couldn't afford to be sick, least of all with such a vague and almost optional ailment.)

Chlorosis was a problem in 1900 and almost unknown by 1920. High school—and especially the elaborate extracurricular and social life that increasingly went with it—is one logical explanation for the disappearance of the disease. A decision by young women to stop dieting to repress their sexuality is another. But the more direct expression of female sexuality began at the lower end of the social scale.

Even as those bemoaning Jazz Age youth were complaining that young women were looking like prostitutes, others noted that the number and visibility of prostitutes were declining sharply. Even as the prohibition of alcohol had created a large illegal industry and increased the power of organized crime and the glamour of vice, prostitution was on the wane. Perhaps there was simply more money in liquor, but it is likely that as society as a whole was becoming more sexualized, there was less business for specialists. While sex for money will never disappear, it continues to play a far smaller role in our society than it did a century ago.

Half a century earlier, during the 1870s, a group of social reformers introduced laws in several states and in cities including New York, Chicago, Philadelphia, and San Francisco, to regulate prostitution. This would, the advocates said, be a boon to public health. In an age without antibiotics, they had a point.

But they lost to a coalition of former antislavery activists, Protestant clergy, advocates for the poor, and women's rights proponents. This loss did not notably reduce prostitution, but it did make a clear statement that while the public was willing to condone prostitution, it was not going to endorse it.

Some who had won this victory, notably members of the Women's Christian Temperance Union, the longest-lived and most effective pro-family advocacy group in U.S. history, decided to go farther. They wanted to actively protect young women by raising the legal age of consent for sex and, therefore, prosecuting more men for statutory rape.

Laws regulating the sexual activity of children and young people were usually only slightly more strict than those regarding their labor. In 1880 the age below which a woman's consent to sexual activities could not be valid was either ten or twelve in most states, following an English common law standard established during the thirteenth century. In Delaware the age of consent was seven.

Those opposed to raising the age of consent argued that such a move would make innocent men prey to unscrupulous women. They painted a picture of men at the mercy of their lusts and easily manipulated by hot-blooded, coldhearted women in their teens. This picture was very different from the strong men and delicate, blossomlike women who were commonly asserted to represent the American norm. *Medical Age*, a Detroit-based magazine, editorialized that raising the law would allow "the licentious, designing *demi mondaine*, many of whom are under the age of eighteen, [to] entice an innocent ignorant schoolboy to her *bagnio* and after seducing him, institute criminal proceedings under protection of the law." When French and Italian appear in the same sentence, it's clear that something nasty is going on.

The fight to raise the age of consent was pursued year by year, and state by state. Some state legislatures were skittish, raising the age in one session and lowering it later. The overall trend, however, was to raise it to fifteen or sixteen years old. At least as a legal distinction, the difference between nice girls and working girls was reduced.

During the 1990s, there has been a flurry of interest in enforcing age of consent laws—"putting the jail back in jailbait" as one advocate put it—as a way to deter teenage pregnancy. Recent studies have shown that about 60 percent of all babies born to teenagers out of wedlock are fathered by adults. Moreover, when the partners are far apart in age, physical force or psychological coercion is frequently a factor. The campaign appears to be politically popular; there is a substantial constituency for "getting tough" on teenage pregnancy.

However, the first case to win widespread attention did not involve a young victim and a much older predator, but rather, an eighteen-year-old Port Washington, Wisconsin, man who impregnated a fifteen-year-old. The young woman said their sex was entirely consensual. The couple intended to marry. Their ages would have raised little concern in the nineteenth century. Even in the 1950s, when the teenage pregnancy rate reached its peak in this century, getting married was considered the appropriate way to deal with it.

In 1997, however, the eighteen-year-old Wisconsin man was arrested for "sexual assault on a child." He was found guilty by a jury, several of whose members said they were in tears casting their guilty vote, but did so because they had no choice. (A majority of the members of the jury petitioned for his pardon.) Although he was sentenced to two years' probation, rather than a jail sentence of up to forty years, he must still register as a sex offender for the rest of his life.

It shouldn't surprise anyone that such a campaign would lead to what some find to be a troubling outcome. The same thing happened during earlier crackdowns on statutory rape. While older sexual predators unquestionably existed, the great majority of men hauled into court were young workingmen a few years older than the woman involved. When women were allowed to be witnesses in the cases, they more often testified on the side of the accused than for the prosecution. Many of those convicted said they intended to "do the right thing" by marrying the girl. And though it's impossible to say whether the laws deterred significant numbers of young people from having intercourse, there were always plenty of new cases to try.

Then as now, pregnancy was the chief evidence for the crime. Sexual abuse that does not lead to pregnancy frequently goes unreported, and when the young woman complains, especially if the man involved is older and in a position of authority, she is often disbelieved.

The biggest problem, though, is that statutory rape laws rely on an essentially passive view of female sexuality. The middle-class advocates of such laws in the nineteenth century believed that young women wanted to be protected from sexuality, but the women themselves weren't always so sure. While it has recently been argued that all relationships between the sexes are a form of rape, there is little doubt that sometimes young women sincerely want to have sex. Moreover, they may take an aggressive role, or even lie about their age if that's what it takes.

"She asked for it" and "She had it coming to her" are among the oldest excuses for male sexual brutality, and those are not the arguments I'm trying to make. Rather, I am arguing that there has long been a tension between those who seek women's protection and those who seek their liberation, and sexuality is at the core of the conflict. For better or worse, the young women who made the difference in the early twentieth century were the girls who wanted to have fun.

Their first step was to get out of the house, at least some of the time. Around the end of the nineteenth century, job opportunities were increasing for young women of sixteen and older. One important reason was that some jobs long considered appropriate only for men, such as store clerks, secretaries, and "type-writers," were increasingly filled by women, whose wages were lower. Moreover, work in garment and other trades, previously done by women in their homes or in small sweatshops, was now performed in large factories. By 1910, more than 7 million women—nearly a quarter of all U.S. women—had paying jobs, about four times as many as in 1870. In the big cities, the percentages were larger and the women likely to be younger. During the first decade of the century, an estimated 60 percent of all New York women sixteen to twenty had paying jobs.

Meanwhile, domestic service, a traditional women's job that offered little freedom, was a declining area of employment for young white women. (For blacks, the story was different. Demand for black servants in northern cities, along with the railroads' demand for porters, helped accelerate African American migration north.) Overall, the percentage of employed women working as servants declined from 61 percent in 1870 to 26 percent by 1910.

Because their jobs were thought merely temporary until they found a husband, and because they typically lived at home, women didn't earn wages high enough to support themselves. Young women who had jobs nearly always lived at home, where they were treated differently than their wage-earning brothers. Usually, the young women were required to turn over their whole pay envelope to their families. Meanwhile, young men the same age typically paid board to their families and kept as much as half of their wages. Young women were also expected to help after work with housecleaning and taking care of their younger brothers and sisters. It was often the young woman's pay that enabled families that were at the edge of middle-class status to send their sons to high school or college.

It sounds pretty grim, but some things made it less so. For one thing, the work week was shrinking in factories from twelve to fourteen hours a day to nine or ten hours. And at workplaces that employed large numbers of young women, a working girl culture was emerging, often based on gossip and speculation that observers found to be sexually frank, or even scandalous. There was even one floor of Macy's department store where, a male observer noted, "There was enough indecent talk to ruin any girl in her teens."

Perhaps the most important thing was that the young women were out in public for much of their time. They were presenting themselves to the public in a way that was very different from either the ideal of the frail, sheltered woman, residing behind many layers of draperies and swags, or the brazen prostitute. The working girls were respectable, but they were most definitely on display. It became important to dress well, to impress their coworkers, their bosses, and in certain jobs, the public at large. As a young shop girl told an interviewer: "One of the employers told me, on a $6.50 wage, he didn't care where I get my clothes from, as long as I have them, to be dressed to suit him." This helped give rise to the practice of accepting gifts from men, many of whom expected something in return.

The working girls encountered coworkers who came from other ethnic groups, with different values and traditions. Still, there wasn't any major ethnic group, either immigrant or native-born, that approved of this sort of behavior by their daughters. Demure behavior, modesty, and virginity at marriage were expected of young women of English stock, as they were of the Irish, Jews, Italians, and nearly everyone else. Yet, when young women started going out by themselves, staying out till late at night, dressing provocatively, and sassing back, their immigrant parents saw it as part of their children falling under the influence of "American" ways. So they were, but these were American ways that the daughters had a hand in creating.

"If I went out and I knew that I'd get hit if I came in at twelve," said one young woman of Irish parentage, "I'd stay out till one." The girls working in the shops, factories, and offices were talking to one another and, worse, listening to each other, rather than to their mothers. Tensions between the generations were on the increase as parents found their daughters' behavior was following standards that were altogether foreign to them. "An' she's comin' in at two o'clock, me not knowin' where she's been," bemoaned the mother of a candy factory worker. "Folks will talk, you know, an' it ain't right for a girl."

Ironically, the double standard implicit in this mother's comment encouraged the sort of gray-market prostitution that emerged as a feature of working girls' culture. If young women's employers had paid a living wage, and if young women had been able to hold on to more of what they earned, they might not have been forced to accept men's "treats" in return for sexual favors. They might have had sex anyway, but not because they needed a new hat.

The rise of commercial amusements—what we nowadays somewhat inaccurately term "popular culture"—was part of a fundamental transformation of American life. Older forms of authority, most notably the family, were increasingly challenged by the new power of money.

In one sense, this is a very old story. We have already seen how economic opportunities led eighteenth-century farmers to liberate their sons at a younger age than custom called for, how the early mills caused girls to leave their rural homes, and how the wage system in nineteenth-century factories raised the value of children in relation to adults and made parents dependent on their children. The institution of the family survived all these changes, as it survives today, by bending its traditions and evolving new ones as the times demanded.

Still, the turn-of-the-century challenge to the family was both larger in scope and more radical than those that had come before. Its result was a consumer-driven culture, one in which social life became much more expensive—and social behavior much more free. Paradoxically, the growing role of money and the things it could buy made society more democratic—or at least more inclusive. As those old-fashioned things that couldn't be bought, "character," for instance, and "good breeding," were devalued, the possibilities for participating in the mainstream of American life were broadened. It's easier to make money than to acquire the manners, much less the bloodlines, of an elite. The existence of this consumer culture allows the overwhelming majority of Americans to believe themselves members of the middle class of a society that is essentially classless, though sometimes crass.

Young people, almost by definition, live on the frontiers of social change, and they were the first to understand how consumer culture forced them to change the way they saw themselves. This was particularly true for young women. Their mothers, whether Irish, Italian, Jewish, or Yankee, wanted to maintain their daughters' marriageability by protecting their virtue and reputation. The daughters saw themselves as part of a much more complex society, in which marriage wasn't the only goal. If she worked in a shop, for example, she couldn't afford to look like a "back number."

She had to be every bit as alluring as the product she sold. Sexuality wasn't something to be sold only once, in a marriage settlement. It was currency used every day, albeit in smaller portions, to survive and to prosper. "I've no more virtue to lose," Fitzgerald had his protagonist muse in *This Side of Paradise*. "Just as a cooling pot gives off heat, so all through youth and adolescence we give off calories of virtue. That's what's called ingenuousness." By 1920, even debutantes were radiating their sexuality, but lower-class girls did it first.

Their preeminent arena was the public dance hall, a new sort of institution. It evolved out of balls and dances that were given in most cities of any size, sponsored by various ethnic and neighborhood associations. These began as events attended primarily by people who knew one another, and appealed to people of all ages. The presence of neighbors, relatives, and family members kept them well within a more traditional value system. Gradually, some organizations understood that these dances could be big moneymakers, and they began promoting them to a wider public. By 1900, owners of popular halls were allowing organizations to hold dances for which they sold tickets, provided that the landlords would get liquor receipts. The most popular were those that appealed to young people and weren't chaperoned. The fully commercial, every-night dance hall was an inevitable next step.

According to Kathy Peiss, whose book *Cheap Amusements* traces the rise of commercial entertainment in New York, that city had 500 public dance halls by 1910. They sought to attract unattached women by charging them less for admission and coat checking than they did men. Girls as young as twelve and thirteen were habitués. Both men and women typically went with friends of their own sex, and male pairs, after deciding which girls each wanted, would frequently "break" the female pair, each taking a girl for the evening. This seems to have been slightly more polite than pure "picking up," which happened too. Those who were keeping company reverted to more traditional socializing, or at least chose less dangerous settings. The presence of the girls attracted the young men, who were increasingly forgoing male-only settings like lodge halls and saloons.

The music was ragtime, that highly formal African American genre that combines joy with what sounds, to contemporary ears, like enormous restraint. It was very likely the first instance of white youth appropriating black music in a way that extended their ability to express sexuality and, not incidentally, to enrage their elders. "Our girls will spend hundreds of dollars taking grace lessons," complained one critic, "and as soon as a ragtime piece of music starts up, they will grasp a 'strange' man in any outlandish position that often will put the lowest creature to shame."

Another description, from 1906, of a dance called spieling doesn't sound very erotic: "Julia stands erect with her body rigid as a poker and with her left arm straight out from her shoulder. . . . Barney slouches up to her and bends his back so that he can put his chin on one of Julia's shoulders and she can do the same to him. Then instead of dancing with a free, lissome, gliding step, they pivot and spin, around and around within the smallest circle that can be drawn around them."

The closeness of body contact was something new, and though the stiffness and jerkiness of the movements sound unpleasant, those who thought the young dancers were enacting stylized sex were not imagining things. Just as important, though, these dancers in the early 1900s were creating and discarding the first set of youth-driven fads. New dances came along regularly: the turkey trot, the bunny hug, the grizzly bear, "shaking the shimmy." Not all were prurient, and high school principals and other authorities tried to keep up so they could determine which to allow at dances sponsored by their institutions. But things changed too fast. By the time a dance was familiar and acceptable to a high school principal, the young had moved on to something else.

"To have a high standard of life," Simon Patten, the pioneer theorist of the consumer economy, wrote in 1889, "means to enjoy a pleasure intensely and to tire of it quickly." That's what young people were doing with their dances, and increasingly, they were doing it with their clothes as well. It was becoming more and more expensive merely to show up. They also seemed to be doing it with the other people they met at the dance hall, though some girls were looking for more. "When I'm eighteen or nineteen, I won't care about it anymore," said one young woman. "I'll have a 'friend' then and won't want to go anywhere."

Some undoubtedly did find husbands, though when this scene was new, that was hardly a sure thing. Men still tended to return to traditional ideas about womanhood when it came time to pick a wife. But once you accept the values of the consumer society and buy into its fast-changing panoply of "necessities," it's nearly impossible to stop. Hence the phenomenon of "the charity girl," a young woman who would have sex in return for gifts and good times, and perhaps because she wanted to. "Don't yeh know there ain't no feller goin' t'spend coin on yeh for nothin'?" advised one. "Yeh gotta be a good Indian, kid—we all gotta!"

It would be incorrect to assume that most girls who went to dance halls were "charity girls," or indeed that they were engaging in sexual acts. They were looking for excitement, not necessarily of a sexual kind. They wanted to be out with other young people, to see what they were wearing, observe how they were dancing, and show off their new clothes and dancing skills. Of necessity, they had to look for young men who would "treat." Telling the girls at work that the man paid for everything actually raised a young woman's status. It's not surprising that many men saw this new milieu not as a variation on traditional courting but as a new form of prostitution. Girls institutionalized in reformatories at the time, almost invariably for some sort of sexual transgression, tended to attribute their behavior not to sexual appetite, but rather to a craving for nice things.

The dance halls were commercial, but they still represented a sort of peer culture, in which young people looked to one another for fashion. A far more powerful and centralized force for setting fashion and changing behavior became nearly ubiquitous during the first decade of the century: the movies. By 1905, there were 700 movie theaters in Manhattan alone, nearly all of them storefront-sized nickelodeons. By 1910, weekly movie attendance in New York City was 1.5 million, a figure equal to about a quarter of the population. Many, especially the young, were going more than once a week. A 1911 survey of schoolchildren found that 16 percent said they went to the movies daily, and 67 percent of Cleveland schoolchildren said they went "almost daily."

The movies were an immensely popular social force that were, like ragtime and the new dances, completely beyond the control of traditional cultural authorities. In their early years, their audience was overwhelmingly working-class. The moviemakers were predominantly recent immigrants who were simultaneously learning about American society and shaping it by what they put on the screen. The moviemakers were loyal to and responsive to their working-class audience. "I think I admire most in the world the girls who earn their own living," said Mary Pickford, the first movie star. "I am proud to be one of them."

Almost as soon as the movies arrived, they provoked moral outcry. In 1907 some New York nickelodeon owners were arrested for violations of Sunday closing laws and for creating public disturbances. In 1909 the Board of Censorship of Motion Pictures was established, with the aim of setting national limits on the movies' conduct. (Local censorship influenced filmmakers, though the industry-wide production code didn't become fully operative until 1934.)

Also in 1909 the New York Society for the Prevention of Cruelty to Children issued a study of the "pernicious 'moving picture' abomination." The report argued, "This new form of entertainment has gone far to blast maidenhood. . . . Boys and girls are together in the room darkened while the pictures are on and . . . indecent assaults on the girls follow, often with their acquiescence. Depraved adults with candies and pennies beguile children with the inevitable result. The Society has prosecuted many for leading girls astray through these picture shows, but GOD alone knows how many are leading dissolute lives begun at the 'moving pictures.' "

One popular bit of urban folklore told of a young girl pricked with a poison needle in the darkness of the nickelodeon, then sold into "white slavery." Nobody knew the girl; she was the acquaintance of a friend of a friend.

The only reformer who attempted a supply-side approach to dealing with the power of the movies was Jane Addams, who opened a theater showing "clean" movies at Hull House in Chicago. She discovered that she could not succeed showing films like an early version of *Cinderella*, while the theaters nearby had titles like *The Pirates, The Defrauding Banker, The Adventures of an American Cowboy, An Attack on the Agent*, and *The Car Man's Danger*.

It seems, though, that the movie theaters themselves weren't the immediate menace that the reformers feared. It's true that having large numbers of men and women sitting together in the dark was something new. But the movies, unlike the dance hall, appealed to people of all ages. Indeed, it was the only commercial entertainment that adult working-class women embraced. Young people had to watch what they were doing, because the woman in the next row might tell their mothers.

Still, the movies did make a difference. They bombarded youth with romance. An avid moviegoer could see more melodrama in a week than she might have seen in a lifetime of far more expensive plays. Moreover, the sexuality of the stars was magnified, close up for all to see. There were longing looks, bedroom eyes, impetuous embraces, and kiss after kiss after kiss. Young people knew that it was not real, that it was, after all, only a movie. But from the movies they learned a lot that they hadn't been told by their parents.

The movies took on far greater power during the second decade of the century, when they moved from their storefront, urban working-class roots and into more impressive theaters in every town in the country. The messages the pictures conveyed were often complex, even paradoxical. For example, working girl that she was, Mary Pickford was an embodiment of frilly-clothed, ringlet-haired femininity, the moral opposite of the flappers who appeared on the screen, and the scene, during the 1920s. The Victorian idea of the woman as the frail flower was evident in everything from cliffhanger serials that placed women in jeopardy to the work of the greatest early filmmaker, D. W. Griffith. Lillian Gish may have hinted at steeliness beneath a traditionally feminine exterior, but in Griffith's films, she was constantly in need of male rescuers.

During the 1920s, many of the most popular movies dealt with the phenomenon of the flapper, along with such other aggressive female creatures as the "vamp," and her relationship to the "good girl." This was a struggle many girls were, no doubt, having with themselves. By 1928, when *Our Dancing Daughters* appeared, the fun-loving flapper herself, played by Joan Crawford, was the one with the right values. She stood apart because of her sincerity, her hunger for life. Although the villainess was a purported "good girl," she was actually a conniving, vindictive hypocrite who schemed with her mother to woo Crawford's millionaire lover away.

Fitzgerald himself endorsed Crawford as the definitive flapper, and it's likely that girls learned something from her performance. A contemporary viewer might remember Crawford as the self-described "Diane the dangerous," ripping off her outer dress and doing a mad dance that ends with her jumping off a platform into the arms of six pomaded young men in black ties. (She was moshing before its time.) Or you might remember the moment when, surprised by the length and passion of her first kiss from the man she loves, she exclaims, "What a service station you turned out to be!" But Diane, though huntress, is also proudly a virgin, one whose wild joy attracts the men but whose well-practiced, subtle rebuffs keep them from getting close enough to threaten her chastity. Young women during the 1920s may have been radiating their calories of virtue, but Fitzgerald notwithstanding, the lesson Crawford taught was more likely the importance of keeping some in reserve.

"In Joan Crawford the true spirit of the younger generation was shown," a seventeen-year-old female told Herbert Blumer, who in about 1930 interviewed young people on the effect movies had on them. "She even lost her man and in the eyes of the older generations they think that when a modern young miss wants her man back she'd even be a cutthroat, but Joan Crawford showed that even in a crisis like that she was sport enough to play fair. And 'play fair' is really the motto of the better class of young Americans, and even in the best products there is always a blemish so why must the younger generation be so shamefully thought of."

"Goodness knows, you learn plenty about love from the movies," said another young woman who had earlier told Blumer that, beyond clothes and furniture, movies didn't affect her fantasies. "You do see how the gold digger systematically gets the poor fish in tow. You see how the sleek-haired, long-earringed, languid-eyed siren lands the men. You meet the flapper, the good girl, 'n' all the feminine types, and their little tricks of the trade."

Did the movies make girls easier? One young African American man thought so. "These actors stirred within me a desire to do an ardent love scene with a girl," he told Blumer. "The first girl that I tried this on said that I was crazy. The second girl wasn't interested. But the third girl actually thought that I really meant what I was saying about her eyes and lips and she permitted me to try out everything that I had planned and this occasion proved successful in more ways than one."

In interview after interview, Blumer's interviewees declared that the movies had little impact on their fantasies or on the way they behaved. They were, however, confident that the movies were the source of reliable information about how other people act. As young people living in a culture where their peers were setting many of the rules, such intelligence was valuable.

They sound self-deluded, and to some extent they were. Nevertheless, their responses point up a very important aspect of the influence of mass media on youth culture and young people's behavior. Young people look to other people their age for confirmation of what the media is telling them. Young people didn't respond to Joan Crawford's flapper because MGM was telling them to, any more than they respond to a song because MTV is showing a video. What young people are doing is paying attention to what is offered them, of course, but to one another most of all.

Those producing the products aimed at the young are also paying attention to them, looking for things they are likely to respond to. Indeed, the direction of youth culture and of commercialized popular culture in general tends to be more a matter of influence trickling upward than downward. It has involved the repeated repackaging of African American music and working-class white lifestyles—often alloyed with an outsider's view of upper-class glamour. When a movie, a kind of music, a fashion, or an attitude strikes a chord, the media are able to amplify it and certify its acceptance. But an individual will accept it only if she knows her friends accept it, and they won't accept it unless it's something they think their friends will accept.

This is a tortured relationship, to be sure. Nevertheless, popular culture as we know it is close to inconceivable without the participation of the novelty-seeking, perennially insecure young.

The 1920s was the first great era of youth fads. Some of them—baggy, tapered-leg Valentino pants, for instance—began in the movies. Raccoon coats were another male fashion. Men seemed to be bulking up, while women were stripping down. The yellow rain slickers that were in fashion with women for a while seem apt expressions of the image of smooth modernity that many women were seeking. Open galoshes for women are harder to explain, except as examples of the young people's frequently perverse tendency to wear clothes in ways that contradict their function. Boys have been wearing backward baseball caps for at least eighty years.

The most important crazes of the era, then as now, involved dancing and, therefore, music. It began with the shimmy, a survivor of the dance hall era, which gave way in turn to the toddle, collegiate ("nothing intermedgit"), charleston, black bottom, and tango. Except for the last, these were fast, jazz-based dances where the couples often danced apart. Unlike the close-touching maniacal movement of such earlier dances as the turkey trot or the spieling, 1920s dancers were making spectacles of themselves at a slight distance from each other. They were less explicit than the "tough dances" of a decade before, but probably because these weren't primarily working-class dances. Those doing the dances weren't enacting intimacies. Rather, they were expressing energy and radiating more than a few calories of sexuality in the process.

An evening of dancing would not consist exclusively of these frenetic, jazz-based dances. There would be plenty of slower numbers where the partners would hold each other and perhaps do a foxtrot. While parents and social critics might be expected to fret more about close dances, these had tradition on their side. The novelty of the dancing was perhaps even more threatening than its presumed immorality. "Don't permit vulgar cheap jazz music to be played," said a statement issued by the National Dancing Masters Association in 1921. "Such music almost forces dancers to use jerky half-steps, and invites immoral variations."

That same year the *Ladies' Home Journal* launched a campaign to ban jazz music from being played in public. "Jazz originally was the accompaniment of the voodoo dancer, stimulating the half-crazed barbarian to the vilest deeds," wrote Anne Shaw Faulkner in the article that began the campaign. "The weird chant, accompanied by the syncopated rhythm of the voodoo invokers, has also been employed by other barbaric people to stimulate brutality and sensuality. That it has a demoralizing effect upon the human brain has been demonstrated by many scientists."

One interesting aspect of Faulkner's article was the attention she gave to ragtime, which had been the source of moral outrage only a decade before. "Negro ragtime, it must be frankly acknowledged, is one of the most important and distinctively characteristic American expressions to be found in our native music," she wrote, adding that it might become the cornerstone of an "American School of Music." Jazz, "that expression of protest against law and order, that bolshevik element of license striving for expression in music," had no such hope.

The "jazz" to which young people were dancing was an ensemble music, "whiter" in sound and more regular in beat than the great 1920s recordings of such black jazz pioneers as Louis Armstrong, Sidney Bechet, Jelly Roll Morton, and others. It took its cue from African American music, and "Charleston," the quintessential 1920s dance tune, was written by the great black pianist-composer James P. Johnson. As has so often happened, white youth had adopted black American music as its own cultural expression, changing it along the way into something just a bit different. Moreover, composers from immigrant backgrounds, notably George Gershwin and Irving Berlin, took these sources and transformed them. They created a new kind of pop music that was, like the movies, completely outside the established cultural hierarchy and which over-whelmed nearly all the traditional sources of authority.

It's funny to read now what Faulkner and others like her wrote. That's not because the music she so deplored has been supplanted by other sorts of music that are far more rhythmically sexual and disorderly. What makes her funny is not that she was wrong but that she was so utterly irrelevant. Music, like so much else, had become a consumer good. There's no way to stop people from buying what they want. And there's no better way to spur sales than to tell young people that what's being offered is something they shouldn't have.

The greatest fad of the era was, of course, the flapper herself. Where, commentators asked, did she come from? The commercial dance halls certainly seem to be one part of the answer.

So too does World War I. In Europe it was theorized that women had to be more blatantly sexual to compete in a marriage market that was shrunken by the tremendous number of men who died in the trenches. This didn't apply to Americans, though. Still, American soldiers had seen different kinds of women and girls in Europe, and the image of the French *gamine*, the mischievous, boyish-figured but very female waif, was an ingredient of the flapper.

Moreover, there were larger social forces at work, among them the victory of the women's suffrage movement, and even, paradoxically, Pro-hibition, which was also an expression of women's power, if not their liberation.

The falling birthrate, the source of so much trepidation at the turn of the century, combined with a robust postwar economy, allowed par-ents to spend more on each of their children. This resulted in more years in school, the breeding ground for youth culture. It also meant that young people were able to tap into more of their parents' resources to fund the high-consumption lifestyles they were buying into.

Anxiety about the fragility of the female body was on the wane, as physical fitness and such activities as cheerleading began to appear in more and more high schools. (Actual women's sports, however, were rare.)

These forces built upon and reinforced one another. For example, the vogue for more athletic dances demanded bodies that were unfettered by the clothing of an earlier age. That led to the practice of providing checkrooms at school and college dances where the women could leave their corsets before going to the dance floor. Many young women eschewed the unmentionable armament that girded their mothers and adopted a one-piece garment known as a "step-in," though the corset checks persisted throughout the Jazz Age.

Short hair was the most visible possible signal of women's greater freedom, as women at the time frequently noted. Hair is, as we have seen, a perennial expression of generational change, and even in the 1920s, symbolism might have been more important than function. It was probably easier to move and exercise with long hair tied in a bun than it was with the shorter hairdo that featured ear-covering protuberances known as "cootie garages," popular in the early 1920s. (Princess Leia wore them in the *Star Wars* films more than half a century later, without provoking their revival.) Wags at the time said that girls started to cover their ears at the precise moment they started showing their knees. A few years later, though, girls bared their ears as well.

A virtual synonym for "flapper," though it could be applied to men as well, was "modern." It was argued that a new, more open, more direct attitude was a necessary adaptation to a fast-changing world, whose conveniences and challenges were both unprecedented. "The times have made us older and more experienced than you were at our age," wrote Ellen Welles Page, a self-proclaimed flapper, in a 1922 magazine article. "It must be so with each succeeding generation if it is to keep pace with the rapidly advancing and mighty tide of civilization." Echoing G. S. Hall, she said that young people were having to face this challenge at the same time they experienced inner turmoil. "There is no-one to turn to—no-one but the rest of youth—which is as perplexed and troubled with its problems as ourselves."

Hall himself made much the same argument in a rather surprising celebration of the flapper he wrote in 1922. "She has already set fashions in attire, and even in manners, some of which her elders have copied, and have found not only sensible, but rejuvenating," he wrote. "The new liberties she takes with life are contagious, and make us wonder anew whether we have not all been servile to precedent, and slaves to institutions that need to be refitted to human nature, and whether the flapper may not, after all, be the bud of a new and better womanhood."

One other obvious suspect for the change in youthful behavior—if not changing ideas of femininity and femaleness—is the automobile. It came on the scene at almost the same moment as the movies, and it was every bit as sexy. The power and speed of the automobile appeared in posters and advertising as extensions of masculine power. Scantily clad women with windblown hair were often in the picture. These feminine figures were supposed to be allegories, but they suggested possibilities. "Can you think of any temptation today that Jesus didn't have?" an Indiana Sunday school teacher asked his class of 1920s high-school-age students. "Speed!" one young man replied. Anyone who has been a teenager since would have to agree it was a pretty good answer.

The most obvious value of the automobile, though, was as a private space in which to be alone. One young New York gang member boasted in 1915 that if he had a car, "There wouldn't be a virgin left in town." A decade later the chief judge of the juvenile court in Muncie, Indiana, declared, "The automobile has become a house of prostitution on wheels."

There is little doubt that automobiles have played an immense role in the lives of American teenagers; getting a driver's license is probably our culture's most important rite of passage. But while some youths used automobiles to impress girls and get them away for a little fun, young people were escaping their families, identifying with peer groups, and adopting more sexually expressive styles long before most had automobiles. Indeed, the fashions began in the cities among people unlikely to have cars. They could, like the young New Yorker, dream of them, but they didn't have them. The youth culture emerged most forcefully on college campuses, most of which prohibited student ownership of automobiles. Cars eventually were a great aid to the dating and petting social life that emerged after World War I, but they didn't make it happen.

The single most important cause for the rise of the flapper, and of the youth culture of which she was the symbol, was the prolonging of youth itself. For anyone who aspired to a middle-class life—which meant more people all the time—high school was essential, and even college was a possibility.

For ever-growing numbers of young people, the real life of going to work and starting a family was deferred, replaced by a student life, played out almost entirely with people one's own age. Young males and young females, most of them past puberty, met every day at high school. Parents could no longer control their interaction. The central social role once performed by the family had been usurped by the aggressively modernizing institution of the high school. "The high school," wrote

Robert S. and Helen Merrill Lynd in *Middletown*, their classic study of 1920s Muncie, Indiana, "with its athletics, clubs, sororities and fraternities, dances and parties and other 'extracurricular activities,' is a social cosmos unto itself, and about this city within a city, the social life of the intermediate generation centers."

Campaigns for compulsory attendance laws had been notably more successful than those for child labor restrictions. By 1918, thirty states required continuous schooling to sixteen, eight until fifteen, nine until fourteen, and one until twelve. In nearly every case, it was possible to get waivers if employers needed young people to work, and fewer than a third of high-school-age students were actually enrolled in 1920. Still, high school had become the normal activity of the teen years, and students' numbers grew. High school enrollment nearly doubled from 2,200,389 in 1920 to 4,339,422 in 1930. In 1920 students made up 28 percent of youth fourteen to twenty; by 1930 the figure was 47 percent.

Rural youths were underrepresented in high schools, and fewer than one in six African Americans went to high school. (Educators, faced with this statistic, argued that most weren't intelligent enough to benefit.) However, a large population of immigrants' children swelled the enrollment of city high schools and helped change them dramatically. Educators weren't altogether sure they could deal with these students, and they sought to adapt the schools to a new role.

Immigration had declined dramatically. The 1920s was a period of intense assimilation, and the high school was reconceived as a sort of Americanization factory, more concerned with citizenship than with scholarship. The highly influential *Cardinal Principles of Education* issued by the National Education Association in 1918 enumerated seven aims: "health, citizenship, vocation, worthy use of leisure, worthy home membership, ethical character, and command of fundamental processes." Only the last of these might possibly have included reading, writing, and math, which parents probably assumed schooling was for. Oakland, California, High School in 1920 introduced a report card providing grades on open-mindedness, seriousness of purpose, assumption of responsibility, willingness to cooperate, thoroughness, initiative, systematic methods of working, behavior, physical fitness, and prompt and regular attendance. "We are now attempting to educate all the children of all the people, and a considerable proportion of human beings have not been endowed with minds capable of development along the lines of academic scholarship,"

California state superintendent Will C. Wood said at the time. "The schools may develop brains; they cannot supply them."

The high school was in itself a challenge to most of the families of its students. The very goal of Americanization was an assault on immigrant values and traditions. The curriculum's implicit message was that whatever your father and mother have told you is probably wrong. The Lynds observed girls in cooking classes being urged to abandon their mothers' "old-fashioned" methods. And the school had a dean of women whose role was to help students with problems—whether social, religious, or educational—that their mothers could not deal with.

With the high school itself joining the movies, popular magazines, and peer group culture to challenge family authority, parents felt they hardly had a chance. "My daughter of fourteen thinks I am 'cruel' if I don't let her stay out until after eleven," one young mother complained to the Lynds. "I tell her when I was her age I had to be in at nine, and she says, 'Yes, Mother, but that was fifty years ago.' "

As at the high schools, the focus of energy in the colleges and universities of the post–World War I years was almost antiacademic. The dean of women at the University of Wisconsin sounded an early warning of what was to come when she observed "a tidal wave of irresponsible joyousness" sweeping through Madison. During the 1920s, college was considered the most desirable place for a young person to be. University men and coeds were the country's premier style setters. Advertising copywriters assumed correctly that they could sell razor blades by telling the public what brand Yale men used (90 percent Gillette, the ad said). Magazines chronicled what they were wearing, what music they danced to, and how they danced to it. Fitzgerald had set the vogue for the collegiate novel, but there were countless more. Princeton dean Christian Gauss half-echoed the school's famous alumnus when he said the university "has become a kind of glorified playground. It has become the paradise of the young."

The prospective student would learn from novels, magazine articles, college newspapers, or a slightly older friend that what was really important at college was learning how to make your way while getting along with others. This was enacted in fraternities, sororities, and student-run extracurricular activities. The lessons of the fraternity house weren't trivial. They taught young people how to operate within an organization, a valuable skill in an economy increasingly dominated by large corporations.

Higher education was primarily for the privileged during the 1920s. But it was so attractive that those who came of age at that time dreamed of providing it for their children, and they eventually did.

Clearly, the traditional courting procedure, which was based on introductions, requests for invitations to call, and chaperonage, couldn't survive either the years spent in a coeducational high school or the evenings spent dancing or at the movies. Young people responded with a new social invention: dating. This was often accompanied by a semi-scandalous practice: petting. Together, dating and petting provided teen-agers with a way to cope with—and enjoy—the tension between their physical maturity and their social immaturity.

It's hard to believe that we have not yet reached the centennial of the first date. Even for those who view it as quaint or old-fashioned, dating seems like something that must have been around forever. The term "dating" seems to have originated in the late nineteenth century, and it had a tough, illicit overtone. By 1914, however, a women's magazine was describing the "dates" of a sorority girl, and the term had taken on the meaning it has had, more or less, ever since.

A date, as it emerged around World War I, was arranged by the two people involved—the man asked—with no family involvement. It happened outside the home, it usually cost money, and the male was expected to pay. The aspect of the date that seemed most radical to older people during the teens and the twenties was that it implied no commitment on either side, no sign of the seriousness of the relationship. The couple might spend time together alone, something only engaged couples had done before, but a day or two later either or both of the two might be out on a date with different people. A popular girl in the 1920s was one who had lots of dates with many different young men. Especially at a dance, men wanted their dates to be attractive to others who would cut in, and the man would cut in on others. To have to dance with the same partner the entire evening was called "getting stuck" and it was humiliating for both partners. It showed they had made a poor choice.

Dates were an extension of the informal interaction that was an inevitable consequence of coeducation in high schools and colleges. Dating was also an excuse to spend money, and a number of businesses—ice cream parlors, movie theaters, skating rinks, florists—counted on this new market. It was also, in this early version, a way to have company without intimacy.

On the surface, dating suggests that families had lost their power over their daughters' social lives, and that young men now held the upper hand. In fact, it was not quite so simple. Courtship was still policed, but by a different authority: the peers of the people involved. The multiple dating and multiple partnering at dances served to slow the development of exclusive relationships. Those involved, however, saw dating as an extremely competitive activity.

Like the Northwestern University women who created a no-date night to allow time to study, they cooperated only to enable themselves to survive the competition. At many high schools, girls joined in pacts to forswear heavy makeup, short skirts, eyebrow pencils, or silk stockings. At one Wisconsin high school, the boys reportedly formed a secret organization to abolish "sheiks," who they claimed were bringing the school into disrepute. School administrators may have coerced students to make many of these pacts. Still, it's likely that some students were happy to be spared the more exotic and expensive weapons of the attractiveness arms race, at least during school hours.

The peer group's chief enforcement mechanism, though, was gossip. A girl's date would tell his friends what happened, the girl would tell hers, and pretty soon everybody knew. It wasn't advantageous in the long run for a girl to be known as "fast," because, while boys might rush to take her out, she might be excluded from the group. Girls looked for boys who were respectful, but not so "bashful" they didn't try anything at all.

"American boys seek one another's company for everything," wrote Spanish social critic Salvador de Madariaga, "Play and prayer, feast and fast, lesson and leisure—all is arranged in common." What shocked him most was the sharing of sexual intimacy in what he called "the land of 'petting parties.'"

"Petting" is a vague term that sounds almost quaint now. By the end of the 1920s, however, it had come to mean virtually all forms of sexual contact—except coitus. How far did your great-grandparents go? Pretty far, at least as the Kinsey reports of sexual behavior reconstructed it in the 1940s.

Petting parties, involving several couples in the same place, were widely viewed as evidence of the depravity of modern youth. They were also a way for the young people to police each other. Couples could keep an eye on one another to see how far the others were going. Those whose conduct went beyond that of the group would only create scandal for themselves. This is not a system that most parents would design for regulating their children's sex lives, but it was a system nonetheless.

Kinsey's studies, which included many people who came of age in the second and third decades of the century, suggest that those who observed that the society was becoming more sexually active were not imagining things. Nearly all measures of sexual activity, especially outside marriage, rose during that period.

What's most interesting about Kinsey's figures in this context is the insight they give about who petted. Among females, before World War I, 29 percent of those surveyed petted before the age of sixteen, while 43 percent of postwar girls did so. For males, the figure was 41 percent of prewar boys, 51 percent of postwar boys. The most petting was done by those who ended up getting the largest amount of schooling; about 39 percent of males who went on to college petted to orgasm between the ages of sixteen and twenty, while only about 13 percent of their least educated counterparts did so. However, the less educated group was more than twice as likely to have had intercourse as those who finished high school. In other words, many young people were petting younger, but those who did so were less likely to have intercourse. And they were the most likely to realize that they were in a time of their lives which was itself a form of foreplay.

ELEVEN
Dead End Kids

I fear that this generation of youth will be lost.

—ELEANOR ROOSEVELT (1934)

After October 1929, flaming youth became a luxury few could afford.

Daughters didn't stop dancing entirely, but they had to do something else besides, perhaps charge 10 cents a dance. On records, at least, the Jazz Age came to an abrupt halt as recording slowed to a fraction of its pre-Crash volume. One of the top hits of 1931 recounted the story of a young man who "found a million dollar baby in a five and ten cent store." There could still be romance, but it had to happen on the cheap.

By 1929, there was an automobile for every 4.5 Americans, and driving and dating had become a way of life. Three years later, there were surpluses of gasoline, and pundits pronounced that autos had been oversold and the oil industry might never come back. It was no longer an embarrassment for a boy in his late teens to be seen riding a bicycle (though four wheels were still a lot more desirable).

Americans were fascinated by youth no longer. Magazines no longer chronicled their fashions. Hollywood stopped offering titillating glimpses of their sexuality. And when youth returned to the movies in the mid-1930s, the cherubic Mickey Rooney rather than the Dionysiac Joan Crawford was its avatar.

Youth culture was in trouble because young people had no money.

Their parents didn't have surplus cash to spend on them, and it was difficult for young people to find work. Indeed, Congress passed laws barring younger teens from work, while federal regulations effectively blocked older teens from finding employment. There was little for young people to do but to "go on the bum," as a visible minority did, or to enroll in high school, as the majority did. High schools were not ready for these new students. They had more students than they could handle and were facing money problems of their own. Besides, they did not know what to teach them. Nevertheless, it was during the Depression years that high school finally enrolled a majority of the school-age population and moved toward being the universal experience of America's young.

This is the paradox: The austerity of the early 1930s could have killed youth culture, and for a time it appeared that it had. Yet, the enforced separation of young people from the economic mainstream and the emergence of high school as the common experience of young Americans led directly to the emergence of teenagers as we know them today. While the Jazz Age had glamorized and eroticized youth, the Depression and the war that followed did something even more decisive to youth: They bureaucratized it. After World War I, young women *chose* to be flappers. After World War II, they and their male counterparts were teenagers whether they liked it or not. They had no other choice.

The New Deal played a key role in the creation of the teenager. It did so in two ways. First, its policies actively discriminated against young people in the workplace. Later, it created new training and aid programs that dealt with what it defined as youth problems, some of them caused by its own policies.

"It is true that nowadays social and economic problems far outweigh in importance, even among young people, problems involving such personal feelings as love, jealousy, and resentment," Aubrey Williams, the first and only director of the National Youth Administration (NYA), wrote in 1935. "The chief problems of young people, which once were the problems of adolescence, have, as a result of the Depression, become largely the same as the problems of society. And youth's problems won't be solved until society's problems are solved."

Williams was, by most accounts, one of the most idealistic of the New Dealers, and his agency was particularly notable for its commitment both to racial justice and youthful leadership. This statement offers hints about how the status of young people was, paradoxically, both challenged and institutionalized during the Depression.

His job title is one clue. There had never been a National Youth Administration before (and it's unlikely that there will ever be one again). It wasn't created until 1935, after the worst years of the Depression had passed. In creating it, President Franklin D. Roosevelt and his top advisers explicitly acknowledged not only that young people were facing unique pressures in the Depression economy, but that the New Deal's relief efforts were often part of the problem. The government had been striving to shrink the labor force so that heads of families had preference for the jobs that existed. Youth had been classified out of the system, and Williams's agency was a somewhat belated effort to find a role for young people in the society.

Underlying Williams's statement that young people were suffering from the same economic ills as everyone else in the society was an assumption that young people shouldn't be concerned with such serious matters. The alternative, "love, jealousy, and resentment," are hardly frivolous; they're the stuff of tragedy. But they are deeply personal, and not part of the economy. Williams's statement reflects the continuing influence of Hall, and it implicitly recognizes the thoroughly apolitical character of the youth mania of the previous decade. Flapper culture was about "thrills," not programs.

The 1920s had also brought a general narrowing of employment opportunities for young people. The 1920s generation of teenagers was relatively small, which is one reason parents were able to support them for extra years as they attended high school. Even though it wasn't yet true that most young people were in high school, the middle class had accepted this as the norm, one reflected in magazines and newspapers. The expected work of a student in his or her teens was an after-school or summer job, done for "spending money." These were likely to be jobs working as a shop clerk or a waitress, or helping out around a filling station, not full-time work in a warehouse or mill.

Williams and his cohorts accepted this middle-class conception of youth's privileged yet marginal role both because much of their constituency was middle-class and because it was convenient. It was difficult enough to deal with all the full-time breadwinners who were out of work. Besides, labor unions were an important part of the New Deal political coalition. Cutting back on the labor of children and younger teenagers had been part of the labor movement's program even when times were good. In 1934, after decades of struggle, national child labor restrictions went into effect. As for young people, the government's strategy was to head them off at the pass. Young people, the policy went, ought to stay in school, and the government's chief role was to find incentives to keep them there.

Keeping young people unproductive was national policy during most of the Depression decade, and it was a surprisingly uncontroversial one. Finding jobs for men who were married with children was the top priority. During the first two years Franklin D. Roosevelt was president, 1.5 million youths lost their jobs as a direct result of his administration's actions.

Nearly all of these job losses were caused by the National Recovery Administration, an exercise in centralized planning that was ultimately declared unconstitutional by the U.S. Supreme Court. Companies taking part in the program were forbidden to employ workers under the age of eighteen. This regulation devastated those who were out of school and had managed to hang on to their jobs during the first years of the Depression.

Moreover, the New Deal did little to soften the blow to these young people. Most work relief programs forbade the employment of more than one member of a household, which made most unmarried young people ineligible. These rules provided some incentive for young people to marry and set up independent households, but not enough. The marriage rate, which had slid to an all-time low by 1932, rose slightly during the economy's sputtering recoveries, but generally stayed low.

Despite these disadvantages, young people still played an important role in many families' economic survival. According to a mid-Depression study of Maryland youths in their teens and early twenties, 57 percent of those whose fathers were unskilled laborers, 64 percent whose fathers were farmers, and even 15 percent of those whose fathers were professionals depended on their children for support.

In the early years of the Depression, there were no programs at all that benefited school-age teens. The Civilian Conservation Corps, the first and best-known New Deal youth agency, set seventeen as its minimum age. The Federal Emergency Relief Administration funded jobs at colleges that needy students could perform to pay their expenses.

Only when the NYA began in 1935 to fund similar jobs that paid six dollars a month to high school students whose families were on relief was anything done to compensate those sixteen and below for their enforced sacrifice to, as the phrase went, "decongest the job market." Six dollars a month certainly wasn't much for up to twenty hours a week of work, but it did have an impact on school enrollment. Studies of high school dropouts had suggested that one of the major things that caused young people to stop their education was that they didn't have a dress or a pair of shoes to wear. With the proceeds of their six-dollar-a-month jobs, needy students did not have enough to compete with the most popular, better-off students. They might have slipped in the side door of the high school rather than marched proudly up the front steps. But at least they had their modest clothes and a place to go.

The particular burdens that the Depression and government attempts to deal with it had placed on young people was acknowledged, perhaps belatedly, by the creation of the National Youth Administration. But the sense that something was owed to the young was probably less important than the fear that young people would recognize what had happened to them and move to do something about it.

Adults seem always to be at least a little bit frightened by the young. During the 1930s, those who looked beyond our borders had good reason to worry. Young people had played a strong role in the fascist movement that had brought Benito Mussolini to power in Italy. The Hitler Youth, which looked from this side of the Atlantic like the Boy Scouts from hell, was 5.4 million strong by the end of 1936. Hitler Youth members were the first and most acceptable exemplars of a discipline that Nazism promised to bring to a country in chaos.

Throughout the 1930s, politicians and commentators across the ideological spectrum warned of the possibility of a demagogue who could mobilize the unspent energies of America's youth. The young were themselves oblivious to such concerns. Year after year, the overwhelming majority of America's young gave reporters and pollsters an image of complacent, inexplicable optimism. They were having their troubles, they said, but they were getting along, and surely something will turn up eventually.

There were several youth organizations, some with Communist ties, that mounted demonstrations, held youth congresses, and frightened a few politicians. But none of these ever won the support of more than a tiny splinter of the college population, much less the larger and potentially volatile population that was out of school and unemployed, or in school because there were no jobs.

Such organizations provided cautionary examples that were useful to those who wanted to do something about "the youth problem." Youth advocates tended to believe that Americans and their elected officials would respond to the needs of the young only if they posed a threat. The blandness of young people often frustrated those who sought to work in their interest. Youth advocates were like the character in adventure movies who acknowledges that the natives are quiet, then adds forebodingly, "Too quiet."

The ambush they predicted never came. One reason may be that the reformers did get some programs in place that staved it off, though these were, for the most part, small in scale and late in arriving. Another explanation is the one some Muncie residents gave the Lynds when they returned for a second study in 1935: that the Depression had brought families closer together by forcing their members to depend on one another.

A subtext of reformers' concern about youth was that young people had a right to feel upset. They were coming of age in a country that had experienced abject failure. Nearly everything that adult experts had told them about the way the world works had proven bankrupt. They had been herded into schools that didn't know what to do with them. Never was there better reason for what later became known as a generation gap. Young people were suffering and those in authority—in government, in business, in school, in church—had scarcely a clue as to what to do about it. Worse, those things that they had done about it tended to hurt young people disproportionately.

Adults could only ask young people to be patient, a tactic they rarely expect to work. This time, to the amazement of many and the exasperation of some, young people heeded the advice. They really were patient, sullen at times perhaps, but uncannily patient.

For her book *The Lost Generation,* journalist Maxine Davis drove all over the country in 1935, talking to young people, listening for revolution. She tried hard to see in the "resignation without resentment" of the young people a mood similar to what she observed in Germany immediately before the rise of Hitler. What she found, though, were young people who were, as she put it, "unknowing conscripts in an army of outsiders." Indeed, her research confirmed other surveys that showed that large numbers of them believed they would finally get their chance to work when a friend or relative with an "in" was finally in a position to do them a favor. This attitude kept young people from taking advantage of the counseling and job placement opportunities that did exist, though it was probably realistic. Personal connections are always important in job searches—and crucial when jobs are scarce.

The youths Davis interviewed were largely oblivious to the larger dimensions of their predicament. There was small-mindedness as usual, as in the case of a young man in Cleveland who suspected that poor people didn't deserve the relief they were getting. And like the barber's daughter in Chicago frustrated by her own idleness and at odds with her family, the young people saw parents and siblings as a bigger problem than Wall Street speculation, the deflationary policies of the Federal Reserve, or the Smoot-Hawley tariff.

"I'd give anything if I had a job," the barber's daughter said. "I wouldn't live here, you bet, with Ma complaining all the time because we don't have enough money, and Pa sore because we always have the same old potatoes and gravy and bread pudding for dinner, and both of them nagging at my brother Gus because he won't tell where he goes all day long. At least they let *him* go out."

But the most striking and characteristic aspect of the barber's daughter's soliloquy is her quiet confidence in the future. "When times are a little better, and I have a job, I'm going to have a room of my own and plenty of clothes. Not that a girl needs a lot of dresses, you know. Just one or two things with proper accessories." It seemed unlikely that people dreaming of buying the proper accessories would make a revolution.

The closest Davis found to a real threat was the cynicism of "Solly Levin," son of a New York leather tanner who had emigrated from Russia: "My old man thought I ought to work. So I quit school, see? But could I get a job? Could Roosevelt get me a job? You're damn tootin' he didn't. I started lookin'. I rode the boxcars. I got hitches. I tramped on my own dogs. Say, I been lookin' over thirty-eight states for more than two years. I picked cherries in Colorado for six bucks a week. I swept the aisles in a cotton mill for $4.40. Livin'? Don't make me laugh." He ended by telling the author that he had decided to quit trying to work and to lie his way to federal handouts. "You go home and tell Roosevelt that every time the gravy train starts, this baby's gonna be on it."

Traveling as "Solly" did was one of the most visible, and for many, threatening things young people did during the Depression.

A significant percentage of those "on the bum"—hanging out in hobo "jungles," riding the rails, hitchhiking—consisted of people in their teens. Estimates of how many varied widely, but they numbered in the hundreds of thousands. There seemed to be at least one case in every community. The classic case was that of a high school student whose father lost his job. The young person dropped out to help support the family but found nothing. He then heard of a possibility a few hundred miles away. Invariably, this proved a disappointment, but it launched him on a career of chasing other illusory opportunities.

Finally, it led to a life of endless wandering, with jobs some days, and begging and minor thievery in between. In 1940 "Sadie Jones," an eighteen-year-old woman from Maine, told a congressional committee that she had been hitchhiking around the country for several years, usually without work and without a place to sleep. Her travels had taken her to all but two of the then-forty-eight states. (She had missed Oregon and Washington.) The congressmen professed to be shocked that nobody had stopped her at state lines, and the people who had given her rides didn't ask her any questions or try to make her stop what she was doing.

But this sort of thing had been going on for more than a decade, and after 1933, publicly supported shelters and soup kitchens helped make it easier. At least they weren't at home burdening their families, the young hoboes explained to those who asked. They couldn't afford to sit around, waiting for the Depression to end.

"Sadie" did break one common pattern. Most youths from rural areas got only as far as the nearest city, where they would stay, while urban youths tended to move from city to city in perpetual motion. Like many other wandering youths, however, she was drawn to Southern California with its mild weather and aura of Hollywood glamour. This prompted one bureaucrat in 1934 to declare that they were pioneers, going west for opportunity, only to find that the frontier had disappeared.

By the 1930s, the old South was undergoing a more radical transformation than the West, as foreclosures and changing farming methods were threatening the tenant farmer system that had replaced slavery. These changes, along with the chronic poverty of the region, provide the background for one of the most explosive incidents of the 1930s— one that involved young people "on the bum."

On the night of March 25, 1931, eighteen young people, nearly all of them in their teens, sneaked singly and in small groups onto a freight train going from Chattanooga through Huntsville, Alabama, to Memphis.

This was not the happy, trouble-free "Chattanooga Choo-Choo" that Glenn Miller's band made into a monument of youth pop a decade later. This train was bound for a brief racial clash, and for some of the youths, an ordeal that didn't end until years or decades later.

Nine of the youths were black, all males. Nine were white, of whom two were women. Four of the black youths, fourteen-year-old friends from Chattanooga, clung to a tank car, and when the white youths came through, on their way from car to car, some purposely stepped on the black youths' hands. "Look out, white boy!" one of the black youths yelled. "You'll make me fall off!"

"That'd be *too* bad," came the reply. "There'd be one nigger less!" A scuffle ensued, involving all the males on the train. When it was over, all the whites had been thrown off the train except the two young women—and one young man who had been pulled back onto the train by one of the black youths whose hand had been trod upon. The train was stopped in Paint Rock, Alabama, and those who had been on the train were taken to nearby Scottsboro. There, one of the young women said several of the black youths had raped her, and the others had raped her friend. The second young woman corroborated her story, then later recanted. There was an almost complete lack of physical evidence, and the accuser's story was inconsistent. She had recently served a sentence for prostitution and had been seen having sex the night before in the hobo jungle. Nevertheless, the black youths were arrested and tried, and some of them spent many years in jail.

The saga of the Scottsboro boys, as the black youths became known, provided Depression-era newspaper readers with a detailed and rather sordid glimpse of the plight of an American underclass, both black and white. By 1931, so-called hobo jungles had become familiar sights near railroad junctions all over the country. They had been there in the 1920s too, but had gone unnoticed amid the general affluence. With the general economic collapse, the jungles had become more populated, and many more people had reason to fear that that's where they would end up.

Most of those involved in the Scottsboro case would have been in economic trouble even if there hadn't been a Depression. The mother of Victoria Price, the woman who claimed to have been raped, seemed to know what she was doing to earn her living and approved of it. Most of the others were products of southern poverty, where schooling was short and work started early.

"By the time I was ten, I was pulling cotton, plowing and planting and everything a grown man could do I did or my daddy would whup me good," Clarence Norris, one of the Scottsboro boys, wrote in his memoir. "I never could go to school but a couple of days out of a week. I didn't go past the second grade. You didn't bite my daddy's bread and go to school after a certain age."

For white tenants it could be as bad, or worse. Financial pressures on many southern landowners tended to displace white tenants in greater numbers than blacks, whom many landlords preferred. However, many landlords were defaulting on mortgages, throwing the entire tenant system into disarray.

The only thing the whites felt they could rely on was their innate superiority to the blacks. Victoria Price didn't want to go to jail for vagrancy. She was confident that even though she was a known prostitute who had been observed having sex in public in the hobo jungle, she could win sympathy by claiming to have been raped by Negroes. (Nearly two years after the incident, Ruby Bates, the other woman on the train, wrote her boyfriend a letter to reassure him she had had sex with whites, not with the black Scottsboro boys. "i was jazzed," she wrote in the unpunctuated letter, "but those white boys jazzed me i wish those Negroes are not Burnt on account of me. . . .")

At the time, observers of the case were distracted by such immediate issues as the institutionalized racism of the Alabama justice system, or the Communist involvement in the case. As the decade went on, though, it became clear that all those involved, white and black, along with countless others, were on the same train. And they were going nowhere.

Young people like those involved in the Scottsboro incident, who stood at the very bottom of the social ladder, were the target population for one of the first and most kindly remembered of the New Deal programs: the Civilian Conservation Corps (CCC). Established in March 1933, the month Roosevelt took office, it enrolled about 2.5 million young men during the nine years of its existence, averaging about 300,000 at any one time. During this time, it was enrolling larger numbers each year than all the colleges and universities of America combined.

Unmarried men, seventeen to twenty-three years of age, enlisted for six months at a time and worked on a wide range of projects identified by the Interior and Agriculture departments. These included terracing agricultural land to prevent soil erosion, planting trees, fighting forest fires, and most popular of all, building trails and shelters in parks and forests. This recreational development gave the CCC one of the most enduring legacies of all Depression-era programs. They earned a small stipend for themselves, while a larger check went to the men's families at home. When CCC enlistees were surveyed on why they joined, 63 percent said the chief reason was to help their families.

The young men lived a quasimilitary life under the command of military reservists. (This provided work for members of a key political constituency.) Each camp also had an educational program, under the control of unemployed teachers. While most camps offered basic literacy training, the teachers were instructed to allow the enrollees' interests to determine what they taught. There was a selection process, but this did not pose an obstacle to young men who had experienced scrapes with the law. One purpose of the program, after all, was to give young troublemakers something to do. That's one reason the residents of some CCC camps weren't welcomed by the residents of some nearby towns and cities.

Roosevelt had started a similar program in 1931, when he was governor of New York, in which street toughs from New York City rode a train each day to Bear Mountain, where they planted trees. In sending troubled youth into the woods, he was following in the footsteps of Charles Loring Brace and other upper-class nineteenth-century reformers, who believed that removing youths from the city improved both the youths and the city.

The CCC, like several other New Deal youth programs, isolated young people in rural camps, away from the cities and their temptations. But by 1933, it became clear that rural America had problems that were every bit as severe as those of the cities. There was no possibility that the program was going to be able to make up for bad schools or an impoverished upbringing, but at least it did give people a sense of belonging to a larger society.

Participants tended to view things in narrower terms. For many of them, three full meals a day was something close to a miracle; abundant food dominated contemporary accounts and memoirs of CCC service. Many men were impressed by meeting other people very different from themselves and learning to trust them in such dangerous tasks as firefighting. The social integration went only so far, however. The races were completely segregated.

Boarding a bus in New York City for a CCC camp in 1935, Luther C. Wandall, a black enrollee, wrote of his fellows: "Almost without exception they were of a very low level of culture. Such low ideals. Of course many were plainly ignorant or underprivileged, while others were really criminal. They cussed with every breath, stole everything they could lay their hands on, and fought over their food, or over nothing at all."

Once Wandall reached the camp, his attitude changed. There was a piano, New York newspapers, a boxing squad, a baseball team, and an orchestra. The work was healthy, and there were opportunities to learn. "For a man who has no work," he concluded, "I heartily recommend it."

One senses that Wandall would have had reservations about "Jonas," an enthusiastic black CCC enrollee Davis interviewed. A semiliterate son of a southern sharecropper, Jonas's family had been forced from the land when their landlord agreed to participate in a subsidized crop reduction program. "Go home?" he told Davis. "*No ma'am.* The guvvament plowed under the cotton. Now I's awukken for the guvvament."

At the time, and ever since, the CCC has enjoyed a reputation as the most successful character-building program the federal government has ever produced. It also represented an investment in the country's natural resources and recreational facilities that very likely had a payoff even larger than Roosevelt and his appointees claimed at the time. And as things worsened in Europe, the quasimilitary training it offered was thought useful in case a sudden large-scale call-up of forces was required.

One thing it did not do, however, is make its recruits any more successful in the job market. Fairly or not, service in the CCC did not convince employers that an applicant had any skill or character traits that warranted special consideration. The training was, indeed, haphazard. And the skills acquired were of limited use in the current economy. Only the federal government was building irrigation systems and nature trails, while most cities were hoping to be able to rehire the firemen they had already laid off.

There is value in making people feel good about doing useful things, especially when the country is in a Depression. But the value is limited. It was not uncommon for young men to finish their six-month hitch in the CCC and go out confidently to look for a job. Facing frustration at every turn, they would often re-enlist in the CCC, then, after a couple of hitches, return to doing whatever they were doing before. It was a very popular program, but not nearly enough.

*   *   *

The clear message to young people was that society wasn't yet ready to deal with them. High school was the institution it had provided for whiling away the time, and more young people took advantage of it.

The schools had problems of their own. The Depression had brought a precipitous decline in the value of real estate—the tax base of American public education. Increased property prices had underwritten the expansion of schooling in the 1920s—especially of high schools. Decimated real estate values devastated schools' budgets. Building stopped. Teachers' salaries were cut, or as in Chicago, not paid at all for long periods. Many districts closed their schools down when they ran out of money for the year, some as early as February. In the 1933–34 school year, one quarter of American students attended schools with shortened sessions, and hundreds of schools closed altogether.

Nevertheless, high school enrollment during the Depression decade increased by about 50 percent, from 4.3 million students to 6.5 million. More significantly, slightly less than half of those between the ages of fourteen and sixteen were students in 1930, while by 1940, two thirds were in school.

There was no point in dropping out at fourteen or fifteen or even sixteen to take a job if there were no jobs to be had. At least, people said, school kept them off the streets. But because many schools were so overcrowded that they had to switch to double sessions, they didn't keep the young off the streets very efficiently.

This onslaught of students came at a moment when the schools were singularly ill equipped to handle it. Even though the federal government eventually made loans to school districts to help them meet their payrolls, allowing more of them to offer full school years, school boards were under heavy pressure to cut "frills."

In the rhetoric of American school politics, "frills" are things that weren't offered in schools at the time the person speaking was a student. Given that a large majority of Depression-era American adults hadn't finished secondary schooling, many saw high school itself as a frill. Many of the things most commonly cut were precisely the things that, theoretically, at least, seemed to be the most needed during a time of Depression. These included night schools, vocational counseling, and manual training.

The school boards that had eliminated those programs had little choice. The entire school finance system had broken down, and it was years before state legislatures came up with money to patch it up, if not reform it. (The most popular tactic, after 1933, was to allocate to local districts the tax revenues realized by taxing liquor, following the repeal of Prohibition).

The high schools were in a predicament. Not only did they have more students, less money, and fewer teachers, they didn't have any real idea about what to do with these additional students.

Educators worried, as they seem to have been worrying for decades, about these new students and their presumed lack of commitment to the academic subjects that teachers and administrators saw as the core of secondary education. In fact, students who received NYA work-study aid scored slightly higher marks overall than did their classmates. Nevertheless, there probably was a large population of students in the high schools who weren't really committed to being there. Educators saw the rush by students to find jobs whenever a shuttered plant reopened or an employer expanded its workforce as evidence of their new students' lack of seriousness. They seemed not to have considered the likelihood that these dropouts were from families that needed the income.

A Muncie teacher told the Lynds that these new students were "soggy intellectually and socially, and usually nonparticipants in school life." Teachers tended to identify strongly with the upper and professional classes, even if they were not paid accordingly.

In *Elmtown's Youth,* sociologist August B. Hollingshead, who studied a small midwestern city and its environs in 1941, found that the high school denied honors to lower-class students, punished them more harshly, and created an atmosphere that encouraged them to drop out. Although teachers believed that the mere presence of the lower-class youths that were overcrowding the antiquated school building was an indication of democracy, the students were sensitive to the social stratification that began with the teachers and administration. One professional-class girl said that her mother wanted her to take a home economics course, but she refused because it would make her "rate" lower with teachers and fellow students. "You could take typing and shorthand and still rate," she explained. "But if you take a straight commercial course, you couldn't rate."

A working-class boy worried that his parents were working very hard so he could stay in school. "What good is it?" he asked, adding that the teachers didn't seem willing to help him. "If we go to the high school dances, nobody will dance with us," said a working-class girl. "They will dance among themselves and have a good time, but they make us feel as if we're not wanted." Elmtown's high school seemed to be a social whirl, with a dance nearly every weekend, Hollingshead reported, but the majority of students never attended any dances at all.

It's dangerous to extrapolate from the case of one little school, but the condescension of educators toward their additional students is apparent in what they wrote in professional journals and said at meetings. Many despaired at the possibility of teaching anything meaningful to these new charges. Moreover, they were skeptical of proposals to increase vocational training, because such classes required expensive equipment and specialized teachers that would draw resources away from traditional subjects.

Still, the students were there, and likely to remain. "What are we to do with our youth up to the age of eighteen or twenty when the best technical engineers and industrial experts are agreed that they cannot be used in the industry or agriculture of the future?" asked Joy Elmer Morgan of the National Education Association in a 1934 article. The answer, she hinted, was that schools would play an essentially custodial role, filling the students' time but not developing their intellect.

This approach, which later became known as "life adjustment," was summed up for many by the various versions of the so-called John Jones letter, purported to have been written by "an actual high school student." This appeared some time in the mid-1930s and popped up in speeches and editorials for years. One version of the letter complained of time spent on dead languages and kings and queens and asked, "Why not have a course in personality in high school? This should be a course in which boys and girls are taught to dance, talk interestingly, dress with good taste, and get along with one another."

This notion of a curriculum based on teaching young people how to appear upper-middle-class was more likely the fantasy of a teacher than a student. Dressing tastefully required money, which the presumed clientele for this lightweight form of schooling did not have. A course in getting along would have had to have been directed at the upper class in the school that enforced the exclusion of others, rather than at the students on the margins. There were, eventually, more sophisticated rationales for the life adjustment approach, but the desire for a low-cost, low-effort approach to teaching students the schools did not want to deal with underlay all of them.

The educational establishment could not afford to completely ignore these students because it appeared that, for the first time since the demise of the academies, the high schools were about to get competition. The new player on the educational scene was the federal government. Some key members of the Roosevelt Administration were highly skeptical of the high schools' ability offer the majority of young people in their teens what they would need to survive in the future. One of these skeptics may well have been FDR himself, though his public position was ambiguous, and he was reluctant to take on the educational establishment directly.

Nevertheless, the scale of the New Deal's educational experiments was impressive. The CCC, despite its shortcomings, suggested an approach to schooling that would build on the students' interests, skills, and ability to make sound decisions. The NYA elaborated this principle in 600 residential centers where young people who were out of school could train for trades and industries. These did not recruit out of the high schools, but to traditional educators, these residential centers looked a lot like schools—and hence competition. School administrator and teacher organizations lobbied against them throughout their existence. Although these educators had little enthusiasm for teaching working-class students, they had even less love for a competing, federally funded education system that might have done the job better.

By the 1940s, the CCC and NYA combined were spending $400 million a year, an amount equivalent to about one sixth of the combined budgets of all the nation's public school systems. By that time, however, the NYA program had gone in a completely different direction. In 1938 the War Department asked the NYA to help eliminate a shortage of airplane mechanics. Soon almost all NYA programs were training young people to be workers in the weaponry industry. The program survived until 1943, long after youth unemployment had ceased to be a problem.

Students completed these courses by attending intensive classes for ten forty-hour weeks. While this program did not inspire the affection some still feel for the CCC, it successfully trained young people in a very short time to perform skilled industrial jobs. It did not, however, provide an alternative to the high school in soaking up the years of youth.

Indeed, the NYA war industries training program demonstrated something very different. It showed that if young people know that they are gaining skills that will be immediately useful, and jobs are available, they are able to learn them very quickly. They didn't need to spend years in technical and vocational schools, as the NYA's advocates had originally believed.

Lengthening schooling may have served to shrink the workforce and maintain breadwinners' wages. But it was no favor to a sizable part of the teenage population for whom high school was an obstacle, and who simply wanted and needed to get on with their lives.

Many of those who learned these complex technical jobs so rapidly were people the public schools had despaired of teaching. Dropouts were a large proportion of those involved in NYA programs. It's tempting to speculate that this rapid assimilation to industrial tasks was aided by the emergence at the time of a new kind of young man's car culture.

For teenagers during the 1920s, the proliferation of automobiles first affected those living on farms, and then members of the urban middle class. Cars were important for dating, both as major prestige enhancers and as zones of relative privacy. The automobile was nearly always a family possession to which the young man had temporary access.

The youthful car culture that emerged during the 1930s was, by contrast, defined not by the usefulness of the car for socializing, but rather by command of its technology. In the early years of the Depression, journalists frequently remarked about the large numbers of young men hanging around automobile repair garages, talking among themselves, complaining that there was nothing to do. Still, they were watching what was going on and learning the technology.

Automobile technology was analogous to personal computer technology today. Despite the automobile industry's setbacks in the early years of the Depression, it was still the most dynamic area of the economy. Cars and trucks were changing the ways people did nearly everything. (Radio, another fascination for many young men, was the only other thing that came close.) Moreover, the car contained the same sorts of parts and processes found in countless other things, from a refrigerator to a battleship. If you accept the psychological notion that young people seek competency, it makes sense that young people who had trouble learning the abstract skills taught in the classroom would want to develop the mechanical skills that made the world work. Hanging around the filling station may well have been a more effective educational experience than going to high school.

Only in the latter part of the 1930s, with increasing prosperity and the very beginning of a wartime industrial buildup, were very many young people able to act on their desires. As Hollingshead noted, these were young men at or near the bottom in both income and social status. They typically bought a jalopy at least ten years old, which they frequently had to tow to the place where they would work on it. Often with the help of friends, they would rebuild the car over a period of weeks or months. The result was virtually a new creation. Then the rebuilder, with a few friends, would cruise over to the next town, drive down the street, or stop at a roadhouse or a dance hall, where they would try to pick up some girls. If they actually got the girls, it was a bonus. The important thing seemed to be the car and the ride with their friends.

The car was not a high school diploma, but as the film of *The Wizard of Oz* (1938) pointed out, having a diploma wasn't the same thing as having a brain. And the car often represented a greater personal achievement than four aimless years of high school.

The youthful car culture developed skills that could be adapted to the industrial jobs that emerged as the country prepared for war, and for soldiering too, when that became necessary. And that points to a moral: Look carefully at what teenagers do when they're goofing off. They might be learning something.

When family income declined during the Depression, discretionary spending on teenage children's amusement was, understandably, one of the first items to go. Youth culture seemed virtually to disappear.

Youth culture is sometimes creative, pioneering new forms of expression—doo-wop music in the late 1940s, for example, or rap in the late 1970s. But it is more often a set of decisions about what to wear, what to listen to—bobbed hair and jazz in the 1920s, flowing tresses and British interpreters of Chicago-style blues in the 1960s. Such choices cost money. Depression-era youth didn't have the economic power to define style as they had in the 1920s, nor, except for the automobile subculture just discussed, was it possible to find young people creating new forms of cultural expression.

Commercial popular culture was no longer synonymous with youth culture. It was, nevertheless, more powerful than ever, as radio broadcast the latest bands. It brought performers right into people's living rooms and turned them into "personalities." The Hollywood studio system, created during the 1920s, reached a kind of perfection, turning out vast quantities of films in a wide variety of genres. They appealed to a wide variety of tastes, but they didn't target specific age groups. Everybody went to the movies.

Movies and, to an even greater extent, radio helped to counteract the greater awareness of class distinctions that had become apparent during the Depression. Popular entertainment was no longer an outsider activity. It was a major industry that had won the largest audience in history. When Americans went to the movies or listened to the radio, they had the sense that the country was talking to itself. It wasn't that everyone dressed as movie stars did, or that they talked as they did on NBC. Still, moviegoers sensed that however glamorous the figures on-screen might be, they had experiences and emotions much like their own.

Young people no longer drove popular culture, but they were still important participants in it. They still danced, though the ways in which they did so provoked fewer adult condemnations. Jazz music, which had been so dangerous a decade before, had become mainstream popular music. There had indeed been a catastrophe, but as far as anyone could tell, the music was not to blame. The young people who had grown up with jazz were young adults by now, and they continued to enjoy the music. The musical style that became dominant, large dance bands with choirs of brass and reed instruments, began in the 1920s, though saxophones were becoming increasingly prominent. Hard-charging, blaring saxophones gave the music a driving intensity that evoked the power and speed of the streamlined locomotives of the time. Nobody listening to a 1930s dance band could conjure up fears of primitive voodoo rites. The sound was industrial, though not in the more recent understanding of that term.

The young still dated, though less expensively and less showily. Because those who were old enough to have teenage children of their own had not, for the most part, grown up dating, their daughters and sons had to fight for that right. But now dating was less likely to be with a lot of different people, and one no longer measured success at a dance by the number of partners one had.

"Going steady" was a new term, and a fairly dangerous idea. It was, in some sense, like an old-fashioned engagement, in that the couple felt proprietary rights to one another. Moreover, the couple themselves, rather than their peers, tended to define the rules of the relationship. Yet, it was unlike engagement in that there was no firm, enforceable commitment to marriage. The age of marriage had long been getting later. The Depression forced those who were ready to marry to delay. And younger people who had someone in mind knew that it might be a very long time, if ever, before they could wed.

The combination of a dating style that encouraged intimacy and the remoteness of the possibility of marriage spurred premarital sexual experimentation. Because youthful behavior had ceased to be a public preoccupation, magazines and newspapers no longer carried horrified accounts of young people's sexuality. Still, there is every reason to believe that as much, or more, was happening during the 1930s. Petting parties, which were a group experimentation, complete with (an admittedly fallible) policing system, gave way to a situation in which many young people believed that intercourse was all right, as long as the two loved each other.

While love and romance had always been part of both movies and popular music, both media became markedly more "romantic" than they had been during the more openly energetic 1920s. A 1935 study of motion pictures found that winning another's love was the chief goal of the main characters in 70 percent of movies, marriage for love in 36 percent, and illicit love in 19 percent, and that these three goals accounted for 45 percent of all characters' motivations.

Nobody ever made a similar study of music, but most of the top-selling records of the decade were dance band arrangements of new tunes that now form the core of American standards. To this day, when a band plays a song for slow dancing, chances are better than even it's from the 1930s. (One reason for this enduring quality is, perhaps, that the songs weren't written solely with the young in mind. "Blue Moon," "Night and Day," and "Stormy Weather" are grown-up songs that Depression-era young people also enjoyed.)

Given the movies' 1920s obsession with youth, a category that admittedly included people well into their twenties, it's amazing how thoroughly this category disappeared in the films of the 1930s. There were children, there were young adults, and musicals and comedies had their ingenues, but youth as a social phenomenon was scarcely visible. Until 1937, when Mickey Rooney made the first of the Andy Hardy movies, a high school was scarcely to be seen, much less any suggestion of the enforced idleness or relentless wanderings of many Depression-era youths.

Mickey Rooney is so bizarre a figure that it's hard to imagine him as a model for anything. Yet, he was one of the top box office attractions of his era—he ranked number one in 1939. In the eighteen Andy Hardy movies, along with the "backyard" musicals with Judy Garland and such other films as *Boys Town,* he helped create an inescapable and enduring teenage image. With Deanna Durbin he received a special Oscar in 1938 for his "significant contribution in bringing to the screen the spirit and personification of youth. . . ."

Andy, a judge's son residing in a big house on a tree-lined suburban block, represented a social position that most viewers would find enviable, and possibly worthy of emulation. The most important thing about the movies is that they showed the teen years as part of childhood. Embodied by Rooney, youth had literally shrunk as a problem. He had his crises, of course, but they were a lesser order of magnitude than real adult problems. He had growing pains. He suffered from puppy love. And while there might be some awkward moments in having a judge for a father, there were unquestionable advantages in having so much wisdom so close at hand.

Such youthful stars as Rooney, Garland, Durbin, and a little later, Elizabeth Taylor had a sexual dimension—though the actors themselves seemed scarcely to be aware of it—on the screen. Rooney made *The Courtship of Andy Hardy* in 1942, the same year he married Ava Gardner, the first of his eight wives. These young people were seen as child stars who had grown bigger, and not suffering through an awkward stage. (Rooney himself looked adolescent for decades—he last played Andy Hardy at the age of thirty-eight—then suddenly senescent.) They were, perhaps, an adult's ideal of what people in their teens could be.

For young people, the Garland–Rooney musicals had another message. The stories involved young people who were just bursting with talent, but whom the adult world was unwilling to take seriously. Frustration at not being recognized as having grown-up skills and talents is a timeless teenage theme. At a time when so many of them had to mark time in their lives, it had particular power. "They call us babes in arms" goes the title anthem of *Babes in Arms,* the first of the Garland–Rooney musicals. "We're really babes in armor!" They decide to take matters in their own hands, put on their own show, and demonstrate how good they are. The climax is a production worthy of the megalomaniacal Busby Berkeley—it really was Busby Berkeley—and it's a smash. Here, at least, young people were able to demonstrate that they could do everything they thought they could. Everyone who saw the movies knew they were fantasies, but it was a slightly different fantasy for young people than it was for adults.

Dead End, a 1937 movie based on a Broadway play by Sidney Kingsley, provided another image of youth. The Dead End Kids were a group of loyal, likable, though none-too-bright teenage males who, in this first appearance, idolized a criminal named Baby Face Martin (played by Humphrey Bogart), who serves as their father figure. These young men jumped out of this high-minded picture and became a pop culture franchise. As the Dead End Kids, the East Side Kids, and the Bowery Boys, they made dozens of low-budget, mostly comic movies during the next two decades. Their rough exteriors masked vulnerability and overall decency. While they showed some outlaw qualities in their first couple of movies, they evolved into a tribe of big-city bumpkins. Indeed, the older Leo Gorcey, Huntz Hall, Gabriel Dell, and the rest of the kids grew, the more childlike they became. It was no accident that they were called kids or boys. Even these apparent reform school candidates—their second film took place in one—turned out to be just immature and in need of a little adult supervision.

The success of these youth-oriented movies at the end of the decade coincided with the appearance of more jobs and a little more purchasing power by young people. Another phenomenon of this period was the rise of the comic book. The most enduring of superheroes—Superman and Batman—emerged during the late 1930s. The story the comics told was similar to that of the Garland–Rooney musicals. It concerned people whose innate powers, in the case of Superman, or abilities and command of technology, in the case of Batman, were immense yet unrecognized by the world at large. Yet these heroes were called upon again and again to save the world. With Europe in turmoil and Japan on the march in Asia, it was beginning to look as if the world would soon need saving.

TWELVE
The Teen Age

Today's teenager is a remarkably independent character.
The fact is, he can afford to be.

—Teen marketing pioneer EUGENE GILBERT, *Harper's* (1959)

The roughly two and a half decades from immediately before World War II to the beginning of the war in Vietnam—from the adolescence of Andy Hardy to that of Gidget—might be termed the classic period of the American teenager.

From the moment the word "teenager" was coined, anonymously, during World War II, it described both an incipient social problem and an economic opportunity. Teenagers were the forgotten children of the Depression years and the neglected children of the war years. But they were also seen as style setters, those who could help America escape the pessimism of those who had faced the Depression as adults and help America consume itself into prosperity once the war was over.

At first, the word "teen-ager" had a hyphen in the middle, a sign that the idea was new and might split apart at any time. There were also alternatives, such as "teener." But like such later demographic coinages as "boomer" and "yuppie," the word took hold quickly because it seemed to describe something real. By 1945, the word was almost always spelled "teenager." It was one word describing one new kind of person.

Even during this classic period, teenage influence waxed and waned, emerging strongly during the war, then pushed to the margin during the great postwar spending spree, then reemerging stronger than ever during the 1950s. Still, ever since Americans first started talking about teenagers, they have played a special role in the culture. In contrast to the flapper era, when collegians set the styles, years of bobby-soxers and teeny-boppers saw the flowering of a youth culture, most of whose members shared the common experience of high school. Teenagers had emerged as a significant cultural force, as a major market—and as a threat to order and decency.

While the youth culture of the period is remembered as golden and innocent—dancing to the jukebox on Saturday night, watching other kids dance after school on "American Bandstand"—it had a dark side. Attitudes at the time ranged from concern to near panic over the menace of juvenile delinquency. As we look back, though, much of this fear turns out to be about social change—the arrival of blacks and Hispanics in the cities, the transformation of the South, the increasing income of the working class, and the emergence of this new creature, the teenager, in every town and just about every family in America.

Money plays an apparently paradoxical role in generating youth culture. Because youth culture is, in essence, a series of decisions about personal appearance and entertainment, it can scarcely exist if its members don't have money they can spend as they see fit in ways wholly distinct from how their parents would spend it. That's why youth culture dried up in the Depression years.

But youth culture also depends on young people's isolation from the adult world. If young people had access to adult jobs and responsibilities, they would have more money, but they would become a less distinct market. Such people might influence youth styles, as working-class teen-agers did during the 1950s. But in that case, their glamour derived largely from their distance from the world of high school and homework.

Since the 1920s, people in their teens have usually been at the fringes of the economy. They hold jobs that pay enough to provide spending money for clothes, entertainment, and perhaps a car. But they have been discouraged from aspiring to the kind of economic independence that would make family and high school irrelevant.

The largest source of funding for youth culture is parents. Even though they may be appalled by specific manifestations of youth cul-ture—nose rings, perhaps, or lyrics that say all women are whores—they accept its validity, or at least its inevitability. They remember the humiliation of wearing the wrong thing, the fear of being rejected by other kids. Many parents try to fight values they see expressed in youth culture, and surveys of young people indicate that involved, principled parents are able to do so successfully. Still, nearly all parents must make their peace with at least some aspects of youth culture, and they help support it.

Unlike the parents of high school students in the early 1930s, for whom the dating and dancing scene were unfamiliar and somewhat unsavory innovations, many of the parents of high school students in the early 1940s had experienced 1920s youth culture. This experience may not have made parents like what their children were doing. Still, they were able to identify with their children's behavior in a way that parents who grew up with different patterns of courting and a firmer sense of respectability probably did not.

It is impossible to talk about the midcentury ascendance of teenage culture without talking about World War II. Indeed, the beginning of wartime preparation and the development of armaments industries provided the money and jobs—both to families and young people themselves—that got youth culture started again. As their elders moved up to better-paying, more skilled jobs, young people took some of the more marginal jobs they had left behind. After Pearl Harbor, this process accelerated as men went to war, women went to factories, and young people found all the work they wanted.

This national mobilization was also an occasion for nation-building. World War II set the stage for a country of fewer differences and more similarities, a fertile environment for, among other things, a universal, media-spread youth culture.

"I got out of high school and three days later I was at boot camp in San Diego," recalled one new sailor. "I was an only child, two months away from being eighteen, and very naive, very unsophisticated, sheltered. Suddenly I'm thrown in with people from Oklahoma, Texas, and the Southeast. Some of them had never used a toilet facility, never brushed their teeth. . . . I met guys who had never seen a Jew. They would look upon me as if I were an ogre."

This awkward encounter between the urban Jew and his rural southern bunkmates was one of millions of such meetings of disparate Americans that unquestionably had a profound impact on the nature of the United States after the war.

The most significant thing about his recollection for the purposes of our story, however, is that his country was willing to wait for his services until after he graduated from high school. High school students were automatically deferred, at least up to age nineteen. Many students were impatient with high school and eager to go to war, so they dropped out. But the overwhelming majority stayed in school until they were done, which resulted in a fighting force eighteen years and older. This was not traditional. Men sixteen and seventeen, and a few even younger, had fought in large numbers in every previous war. About 25 percent of all men in the sixteen-to-twenty-year-old age bracket at any time during the war spent time in uniform, which was a significant percentage but historically low.

It's a mark of how entrenched the expectation of high school had become that this high school deferment was accepted without controversy, even as men who had started families and careers faced being drafted. Even though it had been less than a decade since the majority of high-school-age youths actually enrolled in high school, Americans had become accustomed to seeing members of this age group as something less than adults. Forcing them off to war was scarcely considered.

That meant, of course, that they were at home and visible. Some child labor restrictions were waived or relaxed because of manpower shortages. In agricultural areas, schools resumed the practice of shutting down at harvesttime in order to get the crops from the fields. Even quite young boys were working long and late resetting the pins at bowling alleys, though some wondered whether that was really necessary.

"Now, instead of having fun after school," said a junior high school girl, "I take care of children for war workers from 4 P.M. to 11 P.M. . . . I really don't mind doing all these things because it means my brothers and friends will be home sooner. By taking care of these babies, it means that two more mothers will be available for war work."

The first three years of the war brought a tremendous increase in the size of the labor force, including a large number of housewives who entered war work. But people in their teens constituted an almost equally large number. By 1945, about 3 million young people aged fourteen to seventeen had full or part-time work during the school term, about one out of three. Many held down close to full-time jobs that, combined with school, added up to about seventy-two hours a week.

Mostly, these young people took the jobs in agriculture, restaurants, laundries, bakeries, and retail shops that workers had left to take manufacturing jobs, but a significant number also worked in defense plants. Because sixteen- and seventeen-year-olds weren't being drafted, there was a great incentive to drop out of high school and spend a couple of years at a factory job. Pay was relatively high, though many things of interest to this group—notably gasoline and tires—were rationed.

It was a peculiar situation. People who weren't considered old enough to go to war were holding down adult jobs and making adult incomes. Many fathers were off to war, and mothers off to work, and young people were less supervised. Youth canteens were sponsored by the federal government in areas where military bases and defense industries were having a heavy impact. Young people were allowed a large say in the activities at these canteens—mostly dancing and other forms of mixed-sex socializing. The canteens were important centers of the emergent teenage culture, and were inevitably viewed by many adults as a threat to morality.

There are few things more frightening to adults than teenagers with money and without parents, and that's what they were seeing, particularly in big cities with large military installations and defense plants. But there was a certain intensity and, above all, a sexual impatience to be found among the young nearly everywhere. "Knowing that I am going to war, and knowing that I have enlisted in a tough branch of the service," a high school boy told a 1944 interviewer, "I like to get a little fun that can still be had before I go."

"There was a kind of fierceness, almost a desperation, about people meeting each other at that time," recalled a woman who had spent her late teens in Seattle after the war began. "It might sound corny, but there was that attitude of 'I'm going to be shipping out next week, so let's stay up all night and dance.' " Parents may have rightly feared activities more intimate than dancing.

One woman later recalled the year she was fourteen as the year she became addicted to sailors. "Three of my girlfriends and I went to this church, and we would pick and choose who we thought were the best-looking kids in the group and invite them home with us," she said. She said what happened was quite innocent; her mother would play the piano, and she, her girlfriends, and the sailors would sing. "They'd be gone in a few weeks, and then there'd be somebody else." And she would receive and answer love letters from young men in the Pacific whom she didn't always remember too clearly.

The prospect of war intensified romantic relationships between classmates and neighbors, and reversed a long-term trend toward later marriages. Women would give in to their boyfriends' pleas for something to remember, but the price of the sex was often marriage. Premarital pregnancy was still highly disreputable, which was another spur to preinduction marriages. Besides, the army provided modest financial support to the wives of soldiers, a benefit that was more attractive to some women than the soldiers themselves.

Women who made these choices were still considered more or less respectable. Less so were the so-called "victory girls," or "V-girls," as young as twelve or thirteen, the *Ladies' Home Journal* reported, who were willing to provide memorable experiences to as many young men as possible. "The 'fallen' girls have their reasons, or rationalizations, and freely give them to the questioner. It's patriotic to give 'their all' to men who are in service (or about to be); they must take the pleasure of today for fear that they might not have tomorrow; and—besides—they like it."

Every outburst of youth culture has a soundtrack, and in this case it was big band swing. Like at least two previous generations of young people, this group adopted black music, if not necessarily black musicians, as their own.

This energetic, hard-driving jazz-based dance music had been around for a while; Duke Ellington's tune "It Don't Mean a Thing (If It Ain't Got That Swing)" dates from 1931. Popular music changed around 1940 not because the music was new but because the dances were. While the sweetest songs still sold the most records, hard-driving bands with pistoning trombones, saxophones that could sing or swagger, and tight trumpet choirs galvanized the dance scene. The most successful bands offered both the sweet and the swinging, and the most successful of all during this period was that of Glenn Miller.

Miller, who looked more like an accountant than a pop star, burst onto the scene in 1939 with a band that today seems impressive more for its precision than for its expression. But it was greeted as a new sound with a highly danceable rhythm, something that young people were ready for.

Miller's music was to become inextricably linked to World War II in the public mind. Its trainlike sounds, regimented arrangements, and overall sense of organized energy seem to sum up the way Americans transformed themselves to meet the challenge. Miller cemented this identification by leading the Army Air Force Band before disappearing while on duty in France in 1944, an apparent air crash victim.

The other key youth culture figure came from the sweet side of the music. In 1940 Tommy Dorsey, leader of one of the most swinging and successful big bands of the time, hired a young male vocalist who quickly scored big with a top-selling record of "I'll Never Smile Again." Like other big band vocalists, Frank Sinatra took only one chorus of the song, and let the band hold the spotlight, but not for long.

Sinatra had a gift for delivering a song lyric from the very beginning, but that wasn't what first made him a youth culture phenomenon. Rather, his appeal lay in his frailty, his distinctly unfinished quality. He wasn't an adolescent; he was twenty-four in 1940. Still, he seemed to embody distinctly youthful feelings of longing and vulnerability. He had a sexual edge, but he seemed to need some mothering besides.

While Glenn Miller's sound was of an army moving out, Sinatra's was the boy left behind. (He was declared unfit for service due to, of all things, an ear problem.) Sinatra's stardom was cemented during the war, not overseas but on the home front. In 1942 Sinatra played a sold-out one-month engagement with Benny Goodman's orchestra at the Paramount Theater on Times Square. It marked the first known appearance of swooning mobs of teenage girls who screamed ecstatically and at such a volume that they sometimes drowned out the music. Some of them had been hired to get the excitement started, and it worked. Sinatra at the Paramount was a turning point in popular culture. Two years and quite a few hit records later, Sinatra returned to the Paramount, and more than 25,000 teenagers blocked the streets and brought midtown Manhattan to a standstill. In their insistently public rapture, these teenage girls had established Sinatra as the first of a succession of teen idols. And not so incidentally, they had established teenagers as a force to be reckoned with in the culture.

Once Sinatra got going, things were never quite the same for Mickey Rooney. (Sinatra even ended up marrying Ava Gardner, eventually.) While Andy Hardy was a parent's dream of a high school student, Sinatra was many high school students' dream of a man. While teenage heroes are sometimes teenagers themselves, they are more often people who have escaped the restrictions young people feel in their lives.

These emerging teenagers proclaimed that they were no longer children, but they didn't see themselves as grown-ups either. Bobby-soxers embraced clothing that had distinctly juvenile overtones. The girls wore full skirts that swung about dramatically as they jitterbugged. They wore socks, not stockings, and the boys wore letter sweaters that reminded the world of their high school achievements. Their clothing indicated young people's acceptance, and even celebration, of a less-than-adult status in the society. It was informal, fun, anything but serious.

Some people understood, however, that it could be serious business. By 1942, several clothing manufacturers were making clothes bearing labels such as Teentimers, which had been designed and sized with the teenage market in mind. Magazines, particularly those aimed at women, added features aimed specifically at girls in their teens to attract advertising for products aimed at this group. And in 1944 *Seventeen,* the first magazine aimed specifically at this age group, was started. It defined, for the first time, a distinctive teenage market: millions of young people looking for acceptance, competency, fun, and sex and, therefore, the right clothes, cosmetics, clear skin, great shoes, new music, and all the latest things. Unlike in the 1920s, when young people were the style-makers for the country as a whole, youth-only publications like *Seventeen* and its competitors operate on the assumption that it doesn't matter what adults are buying. Teenagers constitute a lucrative world of their own.

The survival and success of *Seventeen* and its later rivals and imitators demonstrate that teenage culture has been a powerful economic force for more than half a century. Yet, the end of the war proved a setback for teenage influence. While Depression-era austerity had parched youth culture and made it less influential, the end of World War II brought a flood of consumption that swamped youth culture. Teenagers didn't disappear, but they weren't as important or as visible as they had been during the war, or would be again during the 1950s and 1960s.

Teenagers were still a distinctive part of the population—going to high school, dancing, buying movie tickets and clothing, along with magazines, records, and numerous other items. There were still products made specifically for them. But they were a far less important share of the market, both because everyone else was buying so much more and because they made up a smaller percentage of the population than they had a few years earlier.

The teenagers of the decade after World War II had been born during the Depression. This was a period of low reproduction rates combined with hardly any immigration. Fewer children were born during the 1930s than during any decade since the Civil War. There are advantages to being part of a small-birth cohort; it means that your parents have more money to spend on you. That's why some termed these postwar teenagers "the luckiest generation," a small group on whom plenty could be lavished.

More than a year before the war ended, magazines were already filled with descriptions of the postwar world. Life was going to return to normal. Women would leave the factories and return to their proper role as housewives. Ex-soldiers would take their place in factories, which would stop manufacturing the tools of war and convert to production of cars, appliances, and even entire houses. There were items that many families had deferred purchasing because of fifteen years of the Depression and the war, and that others felt they deserved because of the sacrifices of the war. Many people feared that an end to war would lead to a resumption of the Depression, and the prediction of a postwar buying spree was intended to be a self-fulfilling prophecy that would stave off another economic downturn.

Although the country never demilitarized as much as was intended, Americans caught up on consumption with a vengeance. American industry created new makes of cars and whole new varieties of appliances—such as the automatic washing machine, the clothes dryer, the dishwasher, and most dramatic of all, television. Encouraged by federal housing policies, nearly all new housing construction was concentrated in suburban areas, something that virtually forced everything to be bought new.

Meanwhile, the emptying of the sharecropper South and the northward migration of black people turned African Americans from a predominantly rural population to a predominantly urban one. More efficient industrial methods were causing low-paying jobs to drop out of the economy, while those who had held them were able to move up to better-paying jobs. And ensconced in their muddy subdivisions, America's housewives were giving birth to the baby boom.

Department store owners loved to see a young housewife come though the door; she was likely to buy everything from a new wardrobe to a bedroom set, a refrigerator, and a stove. By contrast, the high school girl had money to buy a skirt, a sweater, or a pair of shoes. In the face of a seller's market, she was perceived as getting in the way of a real customer. Some short-sighted merchants actually resisted selling items that were of interest to teenagers because they preferred to appeal to the real buyers.

Paradoxically, one of the things that changed merchants' attitudes toward young customers was many teenagers' impatience with the youth culture, and their desire to get married. The age of marriage had declined during World War II, reversing a five-decade trend toward older marriages. But this had been widely assumed to be an aberration, created by the extraordinary circumstances of war. At it turned out, the postwar era proved to be equally extraordinary, with the United States enjoying unprecedented power and prosperity. Jobs that could enable young men to support a family were relatively plentiful, government-subsidized home mortgages allowed the purchase of houses with low down payments, and families were able to help.

The result was that postwar young people married at even younger ages than their wartime counterparts. By 1947, the median age of first marriage for women was 20.5, which means that nearly half of all brides were in their teens. This dropped to 20.3 in 1950 and 20.1 in 1956, the all-time low for twentieth-century America, and well below the level in other developed nations at the time. In 1955 there were 1 million married teenagers, and in 1959 teenage pregnancy reached its modern-day peak. There was little public concern over this phenomenon, however, because nearly all of the mothers were married. (Marriage ages for males declined even more dramatically than for girls, reaching an all-time low of 22.5 in 1956. But the teenage male was not as desirable a consumer as his female counterpart, because he was several years away from marriage, while she was likely on the brink. Besides, she was more likely to be the spouse who decided what to buy.)

Thus, *Seventeen* was able to prove to retailers that this year's purchaser of a poodle skirt would very likely be furnishing an entire household a year or two later. And she might also be establishing lifetime shopping habits. The magazine prospered during the 1950s, with advertisements for furniture and appliances joining those for youthful fashion and cosmetics.

This premarriage market, while it made manufacturers and retailers interested in young people, was really about teenagers participating in a society preoccupied with marrying, settling down, and starting families. It was a trend that was, in many ways, at odds with the distinctive youth culture that arose during World War II. Young people in the immediate postwar years still danced and listened to music. They bought more cars than ever before. They wanted to look good for each other. But they had little direct impact on the dances, the music, the cars, or the fashion of the time. Everyone danced to "The Tennessee Waltz" and listened to novelty records like "Come On-A My House." The war had brought Americans together as never before. Suburbanization had removed them from traditional communities and social networks and suppressed, for a while, awareness of social class. Mass magazines and, increasingly, television addressed the entire nation in exactly the same way. Often, they spoke of youthfulness and the excitement of living in a new world, but such messages applied more to people moving into new houses in new neighborhoods and dealing with new social situations than they did to people in their teens. When everyone feels young, the desires and expressions of those who truly are young get lost.

What had not changed, however, was the belief that teenagers are at a distinctive stage of life, and that high school is their proper habitat. When war broke out in Korea in 1948, the decision was made not only to allow young men to graduate from high school, but also to defer most who went on to higher education. Many were thus able to defer their military obligation into peacetime, while their classmates who were not high achievers or who could not afford college went to war. The Korean War did not markedly affect the lives of most people of high school age. They were seen as too young for one of the most traditional of uses of men in their teens—to serve as cannon fodder.

During the first decade after World War II, teenagers' influence was felt primarily in industries that had lost their hold on the mass market and needed the young people's business. The movie industry, for example, had suffered a one-two punch. In 1948 the U.S. Supreme Court forced the movie studios to divest their theater chains, which undermined the economic viability of the studio system. And at the same time, television began its inexorable rise, which broke adults of the moviegoing habit. Movie attendance plummeted. But teenagers still liked going out on dates and sitting in the dark.

Appealing to teenagers was hardly the first thing studios thought of. They turned to wide-screen, Technicolor extravaganzas to provide an experience that television couldn't match, and even to such novelties as 3-D. These devices sometimes worked, but making movies to which teenagers liked to go on dates was a lot less expensive. Drive-in theaters, which had first appeared in the 1930s, proliferated rapidly because of looser movie distribution, the availability of inexpensive land, and an increasingly auto-borne teenage population. Drive-ins, which were also popular with parents of young children, were denounced from pulpits and by parents as "passion pits." Indeed, they did seem to give rise to a particular sort of drive-in film that didn't really demand to be watched.

A similar thing was happening to radio. Television was rapidly supplanting the network radio shows that had been such a unifying force in the country during the Depression and World War II. Much of the broadcast day was being taken over by disc jockeys, playing the latest records. But the most interesting things were happening late at night, when it became possible for young people glued to radios in their bedrooms to pull in music that wasn't professional, pasteurized pop.

The best-known strain of this late-night music was black rhythm and blues, to be heard on black-oriented stations in Memphis and other cities, and later specifically directed to young people by Alan Freed in Cleveland. The term actually describes a number of different kinds of music—ranging from hard-edged urbanized blues to the sweet, falsetto-laden close harmony of doo-wop music, improvised and refined on urban street corners. Compared with the great black swing bands of an earlier time—Ellington, Basie, Lunceford, Calloway—this music was unsophisticated. But its rudimentary production values gave it an outsiders' edge that was unquestionably exciting, an expression of the uprooting and transplanting of black America.

Also heard on those late-night radios was country and western music from Wheeling, West Virginia; Nashville; and elsewhere. It was this strain of music, with its multiracial roots in the rural South, that evolved into rockabilly and produced the first breakthrough performers of rock and roll. Bill Haley, whose "Rock Around the Clock" is often remembered as the breakthrough rock and roll song, had only recently gotten rid of his cowboy hat. Elvis Presley, with his deep knowledge of both black and white music and his flair for black style, brought the two strains together.

Still, that story is a bit too simple. Consider Mike Leiber and Jerry Stoller, New Yorkers and quintessential late-night dial spinners with no roots at all in the Deep South, who are nevertheless among the greatest and most successful rock composers. At the age of seventeen, they wrote "Hound Dog," which was recorded in a bluesy version by Willie Mae "Big Mama" Thornton. Later, when Presley recorded it, he was perceived as doing a white version of a black song.

Nevertheless, the cultural influence of teenagers remained muted until 1954 or so, when a number of key events seemed to happen at once. Freed moved from Cleveland to New York, where he won national attention for the music he was playing, and the concerts he promoted there were compared with the epochal Sinatra concerts of 1942 and 1944. Elvis Presley burst onto the pop music scene, just as James Dean did in the movies. The doomed trio played by Dean, Natalie Wood, and Sal Mineo in *Rebel Without a Cause* brought flesh and emotion to sociologists' insights that teenagers' families were losing influence to their peers, and that this situation could be dangerous.

It's probably no accident that this outburst of teenage culture came at a moment when the nature of the market for consumer goods, and the way in which businesses understood it, was changing drastically. During the early 1950s, many forecasters predicted that the postwar catch-up in consumption was about to come to an end. Moreover, because of the low Depression-era birthrate, an abnormally small number of people would be getting married and furnishing households. The Depression was still a vivid memory for many. Businessmen understood that the crash in demand for goods during the early 1930s was far more devastating than the Wall Street Crash of 1929. It appeared that a similar demand crash might be looming in 1954 or 1955.

The good news, though, was that real personal income had risen tremendously. There was money around to be spent, if industry could only come up with enough attractive ways to soak it up. Cars became larger and sprouted fins. Houses took on new forms, like the split-level, and included new kinds of rooms for a more sprawling informal sort of living. Even food was reconceived into new convenient forms—such as TV dinners—which carried higher profit margins and embodied a sense of material progress.

In an economy that was forced once again to induce demand, rather than simply responding to it, the teenage market reemerged as a potent force. The man who had the facts on how powerful it could be, and who is frequently credited with shaping it, was Eugene Gilbert. He had been working at it since 1944, when he was eighteen and came up with two insights. One was that teenagers would respond to retailers who cared about what they wanted. The second, probably more important one was that the best way to find out what young people really wanted was to get other teenagers to ask them. (A third, which he had later, was that his young poll takers could themselves influence their subjects to buy his clients' products and make them fashionable. This diminished the scientific validity of his polling, but it increased Gilbert's profits.)

By the time mainstream manufacturers and retailers were willing to pay attention to the young market, Gilbert had 5,000 teenage pollsters throughout the country, gathering information for his corporate clients and providing data for his syndicated column "What Young People Are Thinking," which ran in more than 300 newspapers. He also had a decade's worth of data that showed, for example, that young people had become markedly more conformist since the war, and thus presumably more susceptible to a well-positioned marketing campaign that played on their insecurities and desire to look and act like their peers.

Gilbert was able to tell businessmen that the 1958 purchasing power of teens was $9.5 billion—ten times the total receipts of the movie industry—two thirds of which came from their parents, and the other third from their own earnings. He could tell them that 57 percent of teenagers bought their own sports equipment and records, 40 percent their own clothes, and 36 percent their own shoes.

Moreover, he argued, teenagers provided an additional spark to the economy by urging their parents to indulge themselves with new, more stylish furniture or clothing, or a new car. "The adult market is depression-conscious," Gilbert said in 1958. "Young people have never known a nonprosperous world." In 1959 he attributed the vogue for button-down shirts, Bermuda shorts, "Ivy League" jackets, separates, ballet slippers, and layers of petticoats to teenagers. He added, "Both parents tend to be confused about matters of taste, confused about values, and all too ready to abdicate decisions—whether about cereals, car colors, furniture, or clothing—to sons and daughters who have definite opinions, shared with large numbers of their contemporaries." His tone was mild chastisement. But, in fact, this parental uncertainty was just what he and others promoting this influential and easily influenced market were counting on.

Gilbert showed that anxiety—both parents' about their children and teenagers' about the esteem of their friends—could be powerful allies of the marketer. The 1950s were an anxious time, as the popularization of psychoanalysis and the vogue for tranquilizers attests. At the time, many attributed this anxiety to the threat of Communism and the prospect of nuclear devastation, though people seemed able to compartmentalize those threats.

It seems more likely that people were anxious about things that were closer to home: new larger families, new jobs, new styles of living. We remember the 1950s as a stable time, largely because that stability was so anxiously asserted. The "Father Knows Best" ethos we recall nostalgically was, in fact, merely a veneer of calm over tumult.

★   ★   ★

One symptom of this midcentury anxiety was the recurrent panic over juvenile delinquency. This was first made manifest during World War II, with predictions from the FBI and others that the absent fathers and working mothers would produce unstable children. Statistics suggest that there was a rise in youth crime during the war.

There were also incidents, such as the notorious zoot suit riots in Los Angeles of late May and early June 1943, that grew from fears of an age-old American phenomenon: urban gangs composed of young members of recent immigrant groups. For more than two weeks, first soldiers, and later many civilians as well, attacked and fought with Mexican-American youths, while the police did little to stop the violence.

The first targets were young men who affected the "pachuco" image. This involved a special vocabulary that included a lot of gangster argot, a defiant attitude, and most visibly, the wearing of oversized suits with draped baggy pants and wide lapels. These zoot suits also owed something to gangster style, though they probably owed more to the example of such black performers as Cab Calloway, whose over-the-top outfits blended style and humor. They were parodies of suits. And because a suit is, inevitably, an expression of authority, these distorted suits were an affront to the established social order. The issue wasn't, as some said at the time, that the loose pants were well suited for concealing weapons. Rather, it was seen as evidence that these young people were unwilling to be either assimilated or invisible.

During the riots, soldiers and sailors—men in uniform finished with training and about to ship out—descended on Chicano neighborhoods in taxicabs. In mobs as large as 400, they patrolled the streets and stores and even broke into houses, looking for young men whose zoot suits they could tear off. As this was repeated night after night during early June, the crowds became less selective, and some who weren't wearing zoot suits, and even some who weren't Mexican-American, were attacked.

The zoot suit riots were but one evidence of racial and ethnic tensions that were coming to a head during World War II. The United States had sent a segregated army off to fight the war, and it had interned American citizens of Japanese descent into camps. The reason the zoot suit riots are significant in recounting the history of teenagers, however, is that it was the beginning of a series of conflicts over young people's costumes. The zoot suiters had given a subversive twist to things they'd found in American culture, and they were punished for it. Similarly, the "hood" look of the 1950s and the hippies of the 1960s have served as the focus of adults' fear and anger toward the young.

After the war, the U.S. Senate first took on delinquency as part of hearings on racketeering, then it convened a subcommittee that held hearings on the subject more or less continuously for nearly a decade. And as early as 1953, when the vanguard of the baby boom was still learning the ABC's, the federal government's Children's Bureau was already predicting disaster. With 42 percent more boys and girls entering the ten-to-seventeen age group in coming years, it warned of an increase of 24 percent in car thefts, 19 percent in burglaries, and 7 percent in rapes.

This boomer threat was several years off. People were more concerned with the crisis they believed was happening right then. "Younger and younger children commit more and more serious and violent acts," wrote psychiatrist Fredric Wertham in his much-quoted 1953 book, *Seduction of the Innocent.* "Even psychotic children did not act this way fifteen years ago." Among his examples were an adolescent who tortured a four-year-old "just because I felt like it," a boy of thirteen who committed a "lust murder" on a girl of six, and four boys, aged fourteen to sixteen, who beat the proprietor of a candy store with a hammer, then drove a knife through his head. Others who were concerned about delinquency could cite dozens of other cases, all equally grisly, to prove the point that delinquency was increasing both in numbers and in the brutality of the crimes.

Some of the horror stories were celebrated incidents, such as the man who, while watching the Giants play a baseball game at New York's Polo Grounds in 1950, was shot to death by a fourteen-year-old wielding a rifle he had aimed randomly from a nearby apartment building. Others were of more local interest, or if both victim and culprit were juveniles, reported only in the most general terms. Youth gangs were found in poorer areas of nearly all major cities, and some instances of ganglike violence were following the white middle class into the suburbs. Police statistics in a number of localities suggested that delinquency rates were growing. All contributed to a profile of youth similar to the "superpredator" who emerged in the 1990s.

Oddly, both the 1950s juvenile delinquent and the 1990s superpredator emerged during a time when youth crime figures were low, or even falling. When a boomer-driven rise in crime materialized during the

1960s, however, it provoked little public alarm. Americans had moved on to worrying about other things.

In the 1950s, as in the 1990s, concern about delinquency helped generate the perception that teenage transgressions were on the increase. Then, as always, people in their teens did some terrible things. Moreover, many cities had, as they had for more than a century, serious and intractable problems with youth gangs. But because of the concern about such violence, newspapers tended to play up cases of juvenile crime because it fit certain preconceived notions of the time. Responding in part to the political climate, police departments set up large juvenile units, or even specialized gang and juvenile narcotics squads. Not surprisingly, arrests for such crimes shot up once there was a bureaucracy dedicated solely to apprehending such offenders. While it's very likely true that the majority of those arrested were guilty of violating the law—though often ones that applied only to juveniles—the overall level of crime was surely not increasing as rapidly as the statistics suggested.

If juvenile delinquents weren't on a rampage, what was happening to generate so much public concern? It's impossible to know, but the explanations put forward for the causes of delinquency offer a clue. They fell into three general categories.

The first explanation concerned the disruption of the family. With fathers away at war and mothers often at work, children and young people were seen to have too much unsupervised time and too little parental guidance. "They were six years old in the last war," read the disclaimer at the beginning of *Blackboard Jungle,* the seminal 1955 juvenile delinquency film, which featured "Rock Around the Clock" on its soundtrack. "Father in the army. Mother in a defense plant. No home life. No place to go. . . . Gang leaders have taken the place of parents."

In the immediate postwar era, despite the aggressiveness with which Americans sought to return to normal, there was nevertheless a sense that things were in flux. Men were spending long hours at their jobs. People were living in new neighborhoods where even the trees were saplings, where everything felt tentative and without firm foundations. Most change seemed to be for the better, but it was happening so quickly and so pervasively, it was difficult for parents to be sure they were doing the right thing for their children.

The second set of explanations for youth crime, identified above all with Wertham, looked at one aspect of this changed landscape: the popular media, especially crime and superhero comic books. "Gardening consists largely in protecting plants from blight and weeds," he wrote, "and the same is true of attending to the growth of children. If a plant fails to grow properly because attacked by a pest, only a poor gardener would look for the cause in that plant alone." Wertham, a psychiatrist, argued that psychoanalytical explanations for delinquency were often beside the point, and blamed instead an industry that was profiting by selling children and young people images of violence, brutality, racism, and sexual degradation. Wertham actually succeeded in modestly toning down comic books—a small success that seems to have deflated his movement. His later campaign to reform television fell flat. Still, Wertham seems prescient in his critique of the way in which the media targeted children and made it difficult for parents to shield them from the violent and sexual images they deployed to sell products. While Wertham may have overreached by asserting that reading a comic book leads directly to garroting one's neighbor, his larger point was that constant exposure to vivid images of murder, torture, and rape must have some effect on those who see them, particularly the young.

The third set of explanations viewed youth delinquents as members of a subculture of teenagers who placed themselves in opposition to adult values. This sociological explanation derived from work done by Talcott Parsons in the early 1940s and was developed by Albert K. Cohen in his classic 1955 study, *Delinquent Boys*. Cohen was very careful to say that his analysis applied only to working-class male gang members in cities, and that middle-class or female delinquents were responding to different imperatives, and possibly other sorts of delinquent subcultures. It was not his fault that large numbers of people confused the subculture of gang-based delinquency with the different, far larger subculture of high-school-age youth.

Throughout this book, I've been using the term "youth culture" in a fairly casual way to speak of a succession of subcultures, made up of different sorts of young people at different times who had different effects on society at large. I think there are enough common threads among these subcultures to justify speaking of Lowell factory workers of the nineteenth century or dance hall habitués of the early twentieth as precursors, at least, of what we understand as youth culture. Still, it's worth remembering that the idea of a separate youth culture developed only during the 1940s and 1950s to describe what was a largely novel phenomenon. And for many parents and other adults, it was a threatening one.

The newly realized dream of universal high school education had unexpected consequences. By the 1950s, nearly every young person was being exposed at high school to the middle-class values of planning, organization, and deferring immediate pleasure to achieve worthy, if distant, goals. Yet, adults looked at their children and saw not a blossoming bourgeoisie, but rather an alien culture in their midst. Young men were sporting haircuts that looked like the rear end of a duck, or even wearing it "Apache-style" in a single ridge across the top. (This coiffure was later revived as the Mohawk.) They were wearing blue jeans, a working-class garment. Young women were wearing mascara. All seemed to be speaking a weird gibberish that parents found unintelligible, but assumed, more or less correctly, to be preoccupied with sex.

The purpose of high school was largely to indoctrinate youth with middle-class standards. But by segregating young people with many others their own age, universal high school education gave teenagers the chance to set standards of their own.

Many schools tried to reassert the power of their values by suppressing the symbols of youthful rebellion. They forbade unconventional haircuts and blue jeans, and after student protests had died down, often announced that good discipline had been restored and that delinquency was near zero. Yet, this generally turned out to be only a short-term holding action. The San Antonio school board tried to expel youth culture from the entire city. It proclaimed success, but didn't succeed.

Perhaps, some argued, the infection of middle-class students by working-class dress, street gang attitudes, and Negro music and language demonstrated the democratic experiment of universal secondary education to be a failure. "There arises," said the report of the Midcentury White House Conference on Children, "the possibility that the standards of the lowest classes can reach some of the boys and girls of other social groups."

There's no question that a broader cross section of the society was in high school than had ever been the case before, and it was inevitable that lower-class styles would have an influence on the traditional middle-class students. Indeed, the teenage cultural explosion that occurred in the mid-1950s was almost entirely an expression of working-class style. Marlon Brando was no teenager, but in films like *The Wild One,* he embodied blue-collar freedom and authenticity.

In *Rebel Without a Cause,* James Dean's appeal derives in large part from his hunger for a truer, more intense life than that of his weak, temporizing middle-class parents, and he finds it in lower-class car culture. And Elvis Presley, a lower-class southern white boy who came to stardom fresh from loading a truck, fused hillbilly style with another lower-class expression, that of black rhythm and blues.

The working-class stud, with a pack of Luckies tucked into the up-turned sleeve of his T-shirt, was a singularly intriguing and threatening figure in 1950s teenage culture. He was subversive precisely because he had escaped, indeed, transcended the traditional teenage constraints of high school and parents. He was out in the world and making his own rules. He was not exactly an adult, not like one's parents at least, but he was certainly no child. He was certainly attractive to girls, but he proba-bly had a stronger impact on boys because he provided an alternate way for them to imagine themselves. Working-class fantasy figures reminded young men of powers they felt they had, but weren't free to exercise. Even Elvis was more popular with boys than with girls.

Did the studious boy with the book bag really envy the incipient dropout in the hot rod? Adults feared the answer would be yes, but the boy was more likely selective about what he coveted. He craved the mobility of the hot-rodder, perhaps, and certainly aspects of his style, but not his future. Thus many high school students looked more trans-gressive than they were, something that other students understood. Today the thought of white-bread Pat Boone appropriating the anarchic "Tutti-Frutti" from Little Richard seems a travesty. But Boone was one of the biggest stars of the era, a serious competitor to Elvis in his prime. He was as far as many teenagers at the time were willing to go.

There was a real basis for the 1950s image of the autonomous, working-class teen. Unemployment of out-of-school men in their teens had in-creased substantially since World War II and was a problem in many locales. Yet, there were industries, especially automobile repair, suburban residential construction, and some types of manufacturing, that took on high school dropouts at decent wages. In a sense, such people represented a throwback to an earlier time, when dropping out of high school was viewed as a rational choice, rather than a social transgression.

The female equivalent of the working-class stud was the teenage wife and mother. Motherhood doesn't seem as rebellious an act, as, for exam-ple, riding with a motorcycle gang. Yet, it represented something far more real than high school, an exercise of young women's primordial powers. Marriage, pregnancy, and motherhood, not always in that order, were widely tolerated, if not desirable, ways for young women to leave high school and begin real life. It helped, of course, that the young fathers could still get jobs that paid enough to support a family.

The essence of a subculture is that there are people around you who understand and support your values and behavior—and others you don't expect or even want to understand you. Although the delinquent subculture may have embodied a complete rejection of the values of those in power, the larger high school subculture was far less radical. High schoolers appropriated superficial features of the delinquent subculture, in part to scare the elders, but also to make a distinctive space for themselves.

Adults, aware that the society was becoming increasingly complex and that jobs for unskilled and semiskilled workers were rapidly disappearing from the economy, overwhelmingly supported secondary schooling for all. Voters might occasionally stop a school board from erecting a building they regarded as "a Taj Mahal," but public support for education was probably as strong as it has ever been.

"As an unintended consequence," wrote James S. Coleman in *The Adolescent Society,* a 1961 study of ten Chicago-area high schools, "society is confronted no longer with a set of *individuals* to be educated toward adulthood, but with distinct *social systems,* which offer a united front to overtures made by adult society." And, as Coleman discovered, the values of these systems brought into being by universal high schooling, were frequently at odds with the goal of the schooling, which was presumably education. At the schools Coleman studied, which included big-city, small-city, small-town, wealthy, and working-class suburban schools, academic achievement rarely ranked above third or fourth place as the focus of students' energies. Athletics and socializing ruled.

When they turned a large part of their children's lives over to high school, parents might have thought that they were turning some of their responsibilities over to teachers. But Coleman found that students believed that teachers had almost no influence in their lives. Parents came first, though just barely, while the judgment of students' peers influenced nearly all decisions about how students spent their time and energies, and what they aspired to. Moreover, by comparing students' scores on intelligence tests with their grades, he concluded that, at most of the schools he studied, the adolescent culture had discouraged many of the ablest from achieving. "In other words, in those adolescent societies," Coleman wrote, "those who are seen as the 'intellectuals,' and who come to think of themselves in this way, are not really those of the highest intelligence, but are only those who are willing to work hard at a generally unrewarded activity."

Coleman's statement may carry an unwarranted assumption—that midcentury America wanted its schools to produce intellectuals. When he surveyed parents, they told him that their greatest hope for their high school children was academic success. But when he surveyed the students themselves, they indicated that their parents would be proud of academic achievements, but perhaps even prouder if they proved to be star basketball players or head cheerleaders. The students may have been misreading their parents, but it was just as likely that the parents communicated values to their children that were different from those they expressed to Coleman and his University of Chicago colleagues.

More generally, the students' emphasis on socializing over academics could also be seen as a reasonable adaptation to the larger society as they saw it. Even today, people pay lip service to the increasing complexity of jobs and the amount of knowledge required to do them properly, but success is more often dependent on operating within a group, making yourself respected, and becoming somebody others trust and want to follow. The methods Coleman observed for becoming successful in high school did not precisely correspond to those required to become successful in the world at large, but they probably had a closer correlation than most of what was happening in the classrooms.

Many observers worried that teenagers were excessively conformist, though the real worry was that they were conforming not with what their parents and teachers wanted, but with each other. And they were living at the absolute high-water mark of American big business—a time when legions of junior executives headed for the office in identical baggy gray suits, and when labor solidarity was propelling factory workers into the middle class. A teenager could reasonably conclude that conformity worked pretty well. Coleman saw adolescent society as a "Coney Island mirror" that reflected, and distorted, adult culture. The scariest possibility of all, however, was that adolescent culture was a truer reflection of the society at large than most adults realized.

Perhaps the oddest mirror of all was "American Bandstand," the weekday afternoon dance party program whose national debut in 1957 signaled the arrival of teenage culture in the mainstream. The program, with an impeccably clean-cut, twenty-seven-year-old Dick Clark as host, originated in Philadelphia at a station owned by Walter Annenberg, who was also the owner of *Seventeen*. In fact, the program had been on locally for five years before it began airing nationally on ABC. It was a very inexpensive way for the number-three network, which had a very limited daytime schedule, to fill an hour and a half of airtime. It quickly became a phenomenon.

All that happened was that a group of teenagers danced with each other while others watched from bleachers, and often a rising star would come on and lip-synch a song. Because it was a live, five-day-a-week program, it depended on its young dancers to make it to the studio every day, regardless of the weather, usually by the elevated train that stopped right next to the studio. The young people who were willing to make this daily commitment soon became highly recognized, and if they changed their dance partners, people from all over the country wrote letters demanding to know why. Unknowingly, they had become characters in a rather plotless soap opera.

"American Bandstand" made teenagers and their music acceptable to Middle America by taking the edge off both. The boys had to wear jackets and ties, with neat, though sometimes expressive, haircuts. The girls had to wear dresses or skirts and blouses, often with sweaters covering up their parochial school uniforms. (Round-collared blouses worn under sweaters became a teen fashion, though it began when priests and nuns objected to their students shaking their bodies in front of the nation while bearing the insignia of St. Hubert's or St. Maria Goretti High School.)

And while the program did feature performances by some of the rock originals, such as Chuck Berry, it was even more associated with the production of such pretty-faced teen idols as Frankie Avalon and Fabian. These were performers who were particularly popular with the youngest girls, who, not incidentally, bought the most records.

Still, although the program worked hard to keep itself bland enough for all of America, it still had a distinctly urban look. Many of the regulars were of Italian descent. A handful were black, and they were required to dance only with each other. (When in 1957 Frankie Lymon, the black lead singer of the Teenagers, was seen dancing with a white girl on Alan Freed's television show, the program was canceled.)

They didn't look like juvenile delinquents, though they looked urban and ethnic, and thus faintly exotic to those watching from suburbia. After more than a decade of parents moving to the suburbs for the sake of the children, the look and style of the city kid was becoming romantic. This was a minor rebellion in itself.

Who was watching? Some were younger children (soon to be known as preteens) learning how to be teenagers. And some were teenagers, also learning how to be teenagers. But, according to the ratings services, close to half were adults. They were among those who wrote letters when regular couples broke up, or favorite dancers found new partners. They studied what the kids were wearing and how they were dancing. In the privacy of their homes, they were condoning, even partaking of, a youth scene they were too old to be members of.

The media, which some feared were corrupting youth, had tamed and exploited the threatening adolescent subculture—and together they put on a real nice show.

THIRTEEN
Boom and Aftershocks

You have the feeling that you're just waiting around to
grow up.

—Student quoted by DAVID MALLERY in
High School Students Speak Out (1961)

Ever since the baby boom got under way in 1946, Americans had been
waiting for it to end. When it started, prognosticators expected it to last
a year or two at the most. After three years, people started to wonder
how much longer it would go, and after five years, school boards began
in earnest—and a bit too late—to build the new schools the baby boomers would require. The total of 40 million children born during the first
decade of the baby boom was more than 50 percent higher than the
total of 26 million born during the previous decade, though it wasn't
until the second decade of the boom that the numbers reached their
peak. As the first boomers reached high school, the births continued,
and by 1964, when the first of them were graduating, people had begun
to see the boom as a permanent feature of American life. That's when,
after eighteen years, it stopped.

The very term "baby boom" connotes an explosive phenomenon,
though it lasted for close to two decades and still continues to shape
American life. Its importance was obvious from the moment it started.
While Eugene Gilbert and others were still struggling during the late
1940s and early 1950s to convince businesses to take note of the economic power of teenagers, marketers were already targeting baby boom

toddlers. In 1948 *Time* noted that the population had increased by "2,800,000 more consumers" the previous year, a novel way of speaking of newborns. "Never underestimate the buying power of a child under seven," Frances Horwitch, Miss Frances of television's "Ding Dong School," told a 1954 advertising conference. "He has brand loyalty and the determination to see that his parents purchase the product of his choice." Insidiously, some supermarket chains began to supply kids with their own miniature shopping carts. They had found that parents usually bought the products the children picked out. (Parental backlash eventually brought an end to this practice.) And while the television networks were slow to serve teenagers as a group, boomer-era children had been regaled with "Howdy Doody," "Captain Video," and dozens of other programs—including the "Mickey Mouse Club," which every weekday reminded them that they would be "the leaders of the twenty-first century." Apparently, boomers were expected to wait an awfully long time to lead, consuming merrily as they did so.

Teenagers too were a fairly novel, exciting, and unsettling population, whose economic power was, at the end of the 1950s, only beginning to be understood. When the boomers became teenagers, something important was bound to happen. The first boomers helped to shape that still-contested set of phenomena known as "the sixties." Many young people challenged American institutions, including, eventually, even high schools, to live up to their democratic rhetoric.

"The sixties" and their aftermath have generated libraries worth of documentation, reminiscence, analysis, and polemic, nearly all of which involve young people as heroes, dupes, explorers, bohemians, or barbarians. The "lessons" of the Vietnam War are learned, unlearned, redefined. There is an overlapping literature on sex, drugs, and rock and roll. The civil rights crusade and its metamorphosis into the black power movement had a strong impact on many young people. The Kennedy assassination seemed to leave everything different. The Beatles. The pill. The draft. Eve of Destruction? Ball of Confusion!

It's late August 1968, a line of Chicago policemen are closing in and I'm being pushed through a plate-glass window in the front of the Conrad Hilton Hotel. The memory is in slow motion, a very sixties, *Bonnie and Clyde* sort of thing. Time stretches out, providing an opportunity to contemplate the disaster that's inevitable.

YOUTH WILL MAKE THE REVOLUTION! said a poster I picked off the

ground that day in Chicago's Grant Park. YOUTH WILL MAKE IT AND KEEP
IT. BE YOUNG. BE BOLD. BE BEAUTIFUL. It was a pretty good sales pitch,
perhaps written by a radical with a future on Madison Avenue. We've
been trying to keep our youth ever since. Yet, there wasn't a revolution,
or if there was, it wasn't young people who made it happen.

Still, something important happened, and young people, including
many in their teens, were deeply implicated, and even more deeply
affected. Erik Erikson had theorized that adolescence involves, among
other things, coming to terms with conditions in the society at large.
Your times are part of your identity.

The students at University High School in Los Angeles learned something about their time one afternoon in 1963 when the principal announced that he wanted students to be driven home by their parents. The
Cuban missile crisis was heating up, and he feared a nuclear holocaust was
at hand. He didn't want his students to be incinerated on school buses.
If the world was to end, he didn't want it to happen on school property.

It's a real 1960s story. The high school, the institution to which the
society had decided to entrust nearly all its youth, regularly produced
such dark farce. Out of this confluence of the cosmic and the petty, the
apocalyptic and the ridiculous was the baby boom teenager challenged
to become an adult.

All boomers grew up with a mushroom-shaped cloud over their
futures, and the early boomers learned in first grade how to "duck and
cover" beneath their school desks to protect themselves from nuclear
annihilation. Still, these terrors had coincided with sunny times of increasing economic opportunity and steady technological progress. The
Cold War shaped boomers' experience, but at least for those born during
the first decade of the boom, prosperity was equally important. One of
Gilbert's surveys found that early boomers believed that, within their
lifetimes, there would be continuous material progress—and that nuclear
weapons would put an end to the world as they knew it.

From the 1940s onward, America's corporations were not ashamed
of using the whining and temper tantrums of baby boomer children to
sell their products. And in a larger sense, they counted on the boomers
to keep demand up and prevent the economy from slipping back into
Depression. The bumper crop of kids drove the move to suburbia, to
bigger houses with more specialized rooms and appliances. Other, less
product-oriented institutions also took advantage of parents' hope that

their children would have widened opportunities. Colleges and universities, for example, expanded their facilities and their faculties on the correct assumption that higher education would become part of boomers' lives. A decade before the first boomer attended college, they were running television commercials to bolster political support for expanding public colleges and universities. These advertisements warned of "the closing college door" and asked, "When the children are ready for college, will college be ready for them?"

"Planning" and "organization" were bywords of the post–World War II era. For many, the war demonstrated that large organizations, engineered for peak efficiency, could attack and defeat the most intractable problems. Still, we couldn't afford to be complacent about our ability to come through as we had in World War II. The Soviet Union was depicted as an even larger and even better-organized society, and one that aspired, as Soviet premier Nikita Khrushchev famously threatened, to bury us.

The elementary schools were a front line in the Cold War, and one that such commentators as Admiral Hyman Rickover, the captain of the first nuclear submarine, argued was startlingly weak. What Ivan Knows That Johnny Doesn't, as the title of a bestselling book expressed it, was a matter of vital national interest. The concern became a full-fledged crisis when, in 1957, the Soviets successfully launched *Sputnik,* the first man-made satellite, and followed that with a succession of space triumphs, including the landing of an unmanned craft on the moon. Overnight, science and mathematics courses took an increased priority. Something called the National Defense Education Act provided more federal money for educational institutions from kindergarten to graduate school. Parents who were accustomed to giving their children so much suddenly discovered that the schools on which they had depended might be failing American youth.

It was in this context that the Carnegie Corporation of New York asked James Bryant Conant, a former president of Harvard freshly returned from service on the front lines of the Cold War as ambassador to Germany, to look at high schools. His report, which was published in 1959, appeared at a key moment when school districts throughout the country were engaged in feverish planning of new facilities in anticipation of the arrival of the baby boomers. Unlike Charles R. Eliot, the Harvard president who sixty-five years earlier led an effort to regularize high schools, Conant had no Committee of Ten to placate, or to add authority to his pronouncements. He spoke in a single voice, but he was immensely persuasive at the time, and his conclusions have shaped the experience of high school for everyone since.

As might be expected, he found that high school students weren't working hard enough. Just about everybody finds that. Like Eliot and his colleagues, he sought to solve that problem by increasing the number of courses students were required to take. Specifically, he was concerned that females were almost entirely missing from advanced math and science courses, while males shunned advanced language courses. Requiring five courses a year, plus electives in the arts, would give both male and female students the opportunity to adapt themselves better for a society oriented to high technology and global leadership.

Conant's report was clearly crafted to dampen Cold War hysteria over secondary education. Conant described the comprehensive high school, which teaches a wide array of academic, vocational, and commercial courses, as a uniquely American institution. Such a school, he argued, was better suited to the national goal of equal access and opportunity than the more stratified secondary schooling found in Europe and elsewhere. The problem with American high schools, Conant concluded, was that they were too small. He argued that a graduating class of 100 was the bare minimum size for any school that aspired to have a full program to serve both those who expected to go into college and those who expected to be prepared for work. In practice, Conant's requirements—which included subject-by-subject ability grouping for every course but "Problems of Democracy"—posed scheduling challenges that could be met efficiently only in even larger schools. Still, he observed that 74 percent of all U.S. high schools didn't meet his 100-senior standard, and that 32 percent of all U.S. students attended these smaller high schools.

For many school districts, the solution to this problem was at hand. Indeed, it was in the sixth grade, which at the time of Conant's report, contained half again as many students as the high school classes. In many localities, the sixth-grade enrollment was two or three times that found in high school classes. And the fifth and fourth grades were right behind them and just as large.

Still, Conant's report had an effect even in such places. It established the large high school as desirable, and encouraged districts to operate those rather than to build new, less sprawling high schools. And more dramatically, it greatly accelerated the trend toward consolidation of small school districts into larger regional districts, specifically so that they could provide a high school of sufficient size. Many states rewrote their subsidy formulas to encourage consolidation of districts, and the larger high schools this made possible. There had been a general trend toward consolidation in any event, but Conant's influence greatly accelerated it.

Studies of the effect of school size and student achievement are contradictory and inconclusive. Big schools can't be blamed for the general erosion in standardized test scores and other measures of educational achievement that came after the first few high-achieving years of the baby boom. Television, for example, has always been a suspect in this decline, while others attribute it to the increased number of younger siblings, who they say achieve less than firstborns.

It is possible, of course, to point to well-run large schools and nightmarish smaller ones. Nevertheless, it's difficult not to see some connection between the growth of the large, bureaucratized high school and the student alienation that became increasingly evident as the 1960s went on. The promise of the comprehensive school was exposure to an array of courses and specialized teachers. The price may have been an environment where teachers and students found it very difficult to establish relationships with one another, and increasingly saw themselves as adversaries. Although teachers' influence on students had been declining for years, the large, bureaucratized high school could only strengthen the influence of peer-group culture.

More broadly, Conant's report embodied a quantitative, technocratic way of thinking that was characteristic of the time, and which eventually provoked a serious backlash. This approach was to concentrate on making investments in physical plant and faculty so schools could offer a wide range of course choices at an affordable cost. The student played the familiar role of consumer, finding fulfillment and identity through the choices made from the offerings available. Yet, the student was in a sense also the product, a piece of raw material to which an efficient educational system could add considerable value. Then, as now, additional earning power was the most frequently cited benefit students could derive from their schooling.

It's probably better to see yourself as a supermarket shopper, passively rolling your cart past shelves full of possibilities, than it is to imagine yourself as a machine that's being assembled or a side of beef being butchered. Neither, however, has much to do with the active, imagina-

tive act of educating yourself. Like most secondary school reformers before him, Conant concerned himself with the courses student took, not
their content or quality.

From a distance of four decades, Conant's study seems to be a bit
of a sham. By the time he went to work, it was already too late to
seriously rethink the high school. The first boomers were already in
junior high by the time the report came out. School districts were already
scrambling to expand their high school facilities, and the buildings they
were erecting would be expected to last for a long time. What Conant
provided was a prestigious endorsement of the status quo, a reassurance
to school boards that they were on the right track. Equally important,
his report was at least a gesture to show that educators and school boards
had contemplated the challenges of the baby boom generation. The educational establishment has always valued the illusion of thought.

Baby boomers had been raised to feel a unique sense of entitlement.
Yet, the single most significant fact about baby boomers is that there
were so many of them, all clamoring for attention. And while businesses
were able to supply the Davy Crockett coonskin caps and hula hoops to
feed their fickle enthusiasms, the schools struggled merely to keep up
with the numbers of students that were moving through. It's really not
surprising that all the energy went into engineering the conventional,
rather than searching for new ways to serve and engage the young. While
baby boomers thought they were each special as individuals, they came
of age in a society that saw them as mostly as numbers.

Today critiques of the bureaucratic experience of American education
come primarily from the right, in the guise of a critique of teachers
unions, political correctness, and calls for taxpayer aid to sectarian schools.
In the late 1950s and early 1960s, when this experience was taking shape,
most criticisms came from the opposite end of the political spectrum.

The technocratic reformers like Conant were, after all, responding
to Cold War imperatives for an educated citizenry in a way that had
great bipartisan support. Teachers unions, far from being the cause of
this phenomenon, were, in large part, a response to it, as teachers sought
a voice in the increasingly impersonal schools and relief from the non-
teaching tasks of administration and policing that were taking up more
and more of their time.

As early as 1957, the educator Kenneth Clark wrote of the problem
he called "vestibule adolescence," in which Americans were spending

year after year of their lives in vague preparation for a distant future. He worried that young people would feel "a sense of exclusion in that period of their lives that could be a most creative period" and argued that for a large number of young people, schooling might not be a compelling or even appropriate way to spend so much of their lives.

And in 1959, the same year as Conant's report, Edgar Z. Friedenberg's *The Vanishing Adolescent* appeared. Friedenberg argued that the process of identity formation in adolescence recently described by Erik Erikson was being short-circuited by contemporary culture. Advertising and the mass media were offering young people false identities, decoys that confused them and left their identities forever unresolved and thus ever-responsive to sales pitches that promised completion. Writing at a time when teenagers were being celebrated in the media both for their creativity and their buying power, Friedenberg argued that the teenager was, in fact, a failed adolescent. The teenager hadn't forged an identity through psychic struggle but rather had bought it ready-made. In a 1965 book, *Coming of Age in America,* Friedenberg extended his critique to the high school, in which he saw administrators preoccupied with procedure and public relations, and teachers who were disillusioned or lazy, producing students who were passive, manipulative, cynical, and inclined to cheat or do anything else necessary to simply get through.

But the work that captured the public's imagination, and was actually read by some teenagers, was Paul Goodman's *Growing Up Absurd,* published in 1960. The schools were only one target of Goodman's critique of a society that he argued was not producing jobs worthy of grown-up lives. "The majority of young people are faced with the following alternative," Goodman wrote. "Either society is a benevolently frivolous racket in which they'll manage to boondoggle, though less profitably than the more privileged; or society is serious (and they hope still benevolent enough to support them), but they are useless and hopelessly out."

These radical critiques of the society and its schools had less impact when they came out than they did later in the 1960s. By that time, the Vietnam War, not to mention long experience in large comprehensive high schools and even larger and more impersonal "multiversities," had brought the technocratic Cold War mindset into question, and the schools along with them.

Writers who observed the high schools in the first years the baby boomers occupied them saw a different scene than that chronicled by Coleman in *The Adolescent Society* only a decade before. The football and cheerleader culture endured, but it was no longer the central pageant of high school life, but merely one choice among several. During the decade of the 1960s, U.S. college enrollment roughly doubled, from 3.8 million to 7.5 million. Colleges were enrolling a percentage of the youthful population comparable to that which went to high school half a century before. Students were looking toward college, past high school, which had lost its status and its ability to confer economic advantage.

There was perhaps no character more reviled than the dropout, that misguided, usually lower-class boy (or pregnant girl) who had decided that staying in high school wasn't worth it. Students were bombarded with statistics about how dropouts' income would suffer in comparison with graduates. Most students had not learned enough about statistics in high school to understand that these numbers were usually spurious. That's because they compared dropouts' income with that of all high school graduates, including those who had gone on to college and further professional training. In fact, the income difference between dropouts and those who ended their schooling with high school was rather small. Moreover, the disadvantage was preserved primarily by the refusal of many employers to even consider job applicants who had dropped out of high school. Thus, any opportunity for adolescent experimentation and risk-taking outside of high school was foreclosed, and severely punished. Those who argued that dropping out could, in itself, lead to an educational experience were in a tiny minority.

Nearly everyone who looked at the schools of the 1960s saw a significant population of what might be termed "internal dropouts." These were young people, often highly intelligent ones, who had given up hope that anything meaningful would happen in the classroom. They gave the teachers the answers they expected or conspired in a lowering of standards, on the theory that the teacher couldn't flunk everyone. They valued school primarily as a place where their friends were, and some boys viewed athletics as the only thing that made high school worthwhile. Friedenberg found that athletic coaches were among the most admired teachers because they had little choice but to address the strengths and weaknesses of each individual. Besides, unlike most of high school, sporting events have an immediate outcome.

Teenagers, like other people, often complain that things are worse than they really are. Even the most skeptical observers found classrooms where students were engaged, teachers inspiring, and learning likely. But there was also plenty to suggest that high school was not the universal solution for adolescence.

The true teenage heroes of the late 1950s and early 1960s weren't dropping out of school, even in their own minds. They were fighting to get in.

They were the young people who, following the U.S. Supreme Court's 1954 decision overturning the principle of separate but equal schools, faced the challenge of being the first black students in the high schools and colleges in the South. They were often surprised at the level of hatred they encountered, the vehemence with which classmates and state governors alike told them that they were not welcome. Still, the responsibility fell on their shoulders. There could never be integrated schools unless some students were willing to integrate them. Thus, at a time when most American teenagers were learning that they would not have any impact on society for at least another decade, a handful of young people from a group who historically had the fewest opportunities were protagonists in one of the great moral struggles of the age.

To be chosen to play a role in history is always a mixed blessing, and several of those involved later recalled that their parents were worried and reluctant to let them go ahead. One such teenager told the psychologist Robert Coles that the prospect of his going to the white high school brought forth from his parents accounts of terror and humiliation stretching back several generations. His generation, the parents said, was the first that didn't have to worry about lynching. "They wanted me to be glad I could walk on the sidewalk," he recalled, "because they used to have to move into the gutter in their town when a white man approached them. But I told them that once you walk on a sidewalk, you look in the windows of the stores and restaurants, and you want to go there too. They said, maybe my children, and I said me, so that my children would be the first really free Negroes."

This son argued with his parents, though it was over how best to satisfy a goal that both shared. The parents, understandably, worried over the son's safety, while the son had to come to terms with his historical moment. The way Coles presents this story makes it a demonstration of Erikson's observation that adolescents' struggle for identity necessarily involves understanding and seeing a role for themselves in the context of their own time. That doesn't mean that they're denying history or their parents' values, just that they might see a need to proceed in ways different from those that their parents might choose. Such a process produces change, and it can involve serious disagreements between parents and children, but it doesn't lead to the sort of generation gap that many observers noted as the 1960s proceeded.

There was no such generation gap between the young civil rights pioneers and their parents, nor was one apparent, at least at first, between the white students whose schools were being integrated and their parents. Once the young black students made it past the racist politician who blocked the schoolhouse door, they still had to deal with classmates determined to make life miserable for the interlopers.

The civil rights movement was a wellspring of youthful activism, as students and others, both black and white, began to participate, first in summer organizing campaigns, and later, by calling attention to de facto school segregation and other forms of discrimination in the North. Most of those involved were college students, although there were some of high school age, including several martyrs of the movement.

The Freedom Rides of 1960 and 1961 were organized by the Student Nonviolent Coordinating Committee (SNCC), an organization that initially promised an alliance of black and white young people and later became a vehicle for the black power movement. The Freedom Rides were something new. Americans had spent much of the previous decade bemoaning the conformity and complacency of their young. Now, at least, young people were taking action.

The students who became involved in civil rights and other causes did not, however, come out of nowhere. Many parents of baby boomers were products of the Depression, when the New Deal had brought a concern for social justice into the mainstream. Several studies of activist students concluded that the great majority came from families that approved of their goals, if not always of their tactics. What was emerging was not a generation gap, but rather unresolved issues in American society that had been forgotten during the war and its euphoric aftermath.

Indeed, to the extent that there was a generation gap, it was between parents whose experience had been shaped by scarcity and who worried that a Depression might recur, and children who had known nothing but plenty and expansiveness. Young people are naturally more optimistic about how problems can be solved, because they have not learned how intractable some can be. But the baby boomer generation, which had grown up in a country that commanded a greater percentage of the world's resources than any nation before or since, had strong grounds for its optimism. If, as the country's leaders said, money and organization could accomplish anything—even put a man on the moon—there was no time like the present. David Mallery, a teacher who traveled to high schools throughout the country to research *High School Students Speak Out,* a book on early 1960s students, made a practice of asking those he interviewed what historical period they would choose to live in. A majority surprised him by picking a point in the near future, when space travel and other technological marvels would transform everyday life. Full adulthood was receding farther into the future for them, but they believed that future was going to be great.

Still, in this future orientation they were not very different from their parents. For example, the appeal of John F. Kennedy in the 1960s presidential election was explicitly youthful, and his rhetoric of a New Frontier fused television's western-series mania with the excitement of space exploration and scientific discovery.

One of the most careful studies of the first wave of baby boom teenagers was made by psychiatrist Daniel Offer and his collaborators, who repeatedly interviewed and tested male students at two Chicago suburban high schools. They were seeking to understand the "normal" middle-class teenager, in contrast with most other psychiatric studies that extrapolated the conflicts felt by their adolescent patients to the entire teenage population. They found that nearly all the young men they studied had a period of conflict with their parents, at around the ages of thirteen and fourteen, characterized by bickering and conflicts over rules and dress. But once these squalls had passed, most of those studied became increasingly confident of themselves in almost every way, and they looked to their parents for guidance in the most important decisions of their lives. Peers were important, especially for people who were spending most of their lives in an age-segregated environment, but parents counted for far more.

In *The Psychological World of the Teenager: A Study of Normal Adolescent Boys,* published in 1969 when the generation gap was depicted in the media as huge and unbridgeable, Offer argued that there was no generation gap at all. Indeed, the young men he interviewed told him that many pressures generally believed to come from youth culture—above all, to begin dating—actually came from their parents. Parents were seeking reassurance that their sons would be successful and "normal" in their behavior patterns, so they pushed their children into doing things they didn't really feel ready for. This was a problem mostly for the younger boys. By the time they were older, nearly all wanted to date and didn't feel any conflict. But it does suggest the possibility that parents' worries about their sons' acceptance by the youth culture may lead them to exaggerate its influence and underestimate their own. (Offer's research seems to hint that parents' fears that their sons might be homosexual spurred them to encourage premature heterosexual behavior, but he did not address that directly.)

Offer even went so far as to argue that, while there were youthful costumes and youthful products, there was no such thing as a youth culture. He found that young people strongly reflected the values of their parents, and although they sometimes appeared to speak a different language among themselves, there was strong communication between young people and their parents. But an inevitable gap—between what people say and what they do—along with the perceived gap between generations, certainly appeared to be garbling the conversation.

Offer's data also buttressed others' observations that teenagers did not think highly of their high schools. When asked to name adults who had an influence on them, teachers were rarely mentioned, and guidance counselors and school administrators almost never. There's good reason to believe that the authority figures against which teenagers rebelled in the 1960s weren't their parents but the principals and staff at their high schools.

The first boomers were in their last year of high school on November 22, 1963, when President Kennedy was assassinated. Until then, despite the discontent that was bubbling below the surface, it seemed that the progress of the post–World War II era was continuing, perhaps even accelerating.

Rock and roll had, to be sure, stayed around a lot longer than many had expected it to. Baby boomers endorsed their predecessors' music and developed it in some new directions. Perhaps as an indirect result of the civil rights movement, black performers began to have the hit versions of popular songs—without fear of a cover version from white artists. Many of these popular black performers were females, though the lyrics they sang were usually about devotion to a boy. Only in 1967, with Aretha Franklin's recording of Otis Redding's song "Respect," did black power, women's liberation, and the sexual revolution all come together in a single song. And while white artists had previously had hits by singing black music, a lot of the music performed by the new black groups had been written by white songwriters working out of New York's Brill Building. Berry Gordy's Motown releases were written by black composers, signaling a multiracial American pop. Before long, black groups were covering white groups' songs.

The Beach Boys, perhaps the most creative early 1960s group, appeared to embody a laid-back but high-consuming hedonism that must have warmed the hearts of those hoping to sell to the boomers. The girl who had fun, fun, fun till her daddy took the T-Bird away might still be able to convince her indulgent parent to buy her a Mustang. (The Mustang was developed for a spring 1964 release, just in time to be a high school graduation present for the first class of baby boomers.)

Young people did change the way they danced. The "American Bandstand" dancers, even while dancing to rock and roll, followed the tradition of memorized steps, with the male in the lead, and lots of touching—which previous generations of youth had thought was the whole purpose of dancing in the first place. In the 1960s, dancing became one of the first places where female equality was asserted as the members of the couple, united primarily by eye contact, moved with a freedom and energy that couldn't be accommodated within the social convention of the coordinated couple. There were dance steps, such as the twist, the mashed potato, and the Watusi, but these were more like parodies of dance steps, conventional forms of mime that could be integrated into the dancers' mutual display. Except for the twist, they were more popular with adults trying to dance like teenagers, at clubs like New York's Peppermint Lounge, than with teenagers themselves.

Apparently countering the influence of rock and roll among young people was the folk music revival. This music had its roots in the 1930s radicalism and became associated with civil rights and other protest movements. By the late 1950s and early 1960s, this leftish music seemed less threatening to some than rock and roll. This was particularly true when it was performed by such clean-cut acts as the Kingston Trio or introduced on the television show "Hootenanny" by Jack Linkletter, whose conservative father, Art, had made a career of inducing baby boomer kids to say vaguely salacious things on daytime television.

These sorts of changes fell within the normal range of the dynamics of fashion. They were not disruptive. Indeed, the economy, and especially that part of it that is driven by the young, counts on the regular churning of desires.

But after November 22, 1963, change changed. It happened faster. It went deeper. It felt more real.

If the President could be shot, it seemed, just about anything could happen. The crazed normality of the 1950s and 1960s—where high school principals fretted about hall passes and Armageddon—had been shattered. This event that wasn't supposed to happen called into question all the things that *were* supposed to happen.

Would "the sixties" have happened without the Kennedy assassination? Many who lived through the time would say no, though in most respects, the answer has to be yes.

Surely the Beatles would have arrived on these shores, with their long hair that seems, in retrospect, so shockingly short. "I wanna hold your hand," they sang. Elvis and Jerry Lee clearly had more on their minds than that, but of course being gently sexy, even androgynously sexy, was one of the trends the Beatles helped start. And the indirect message of the Beatles, that youth culture existed outside of America, would certainly have registered with the boomers.

As for the sexual revolution of the 1960s, it was already getting under way. In *Playboy,* Hugh Hefner presented great casual sex as a lifestyle essential, along with fast cars, martinis, and state-of-the-art stereo equipment. The birth control pill came on the market in 1961, making reproductive responsibility invisible to males. A large part of the sexual revolution was franker public talk and increasingly blatant use of sexual imagery in advertising and the movies, something that was already under way before Kennedy's death.

LSD was also in existence before Kennedy's death, even developed by the government. Timothy Leary's notorious slogan, "Turn on, tune in, drop out," would certainly have had an appeal to those internal dropouts that populated so many high schools. But whether "blowing your mind" and exploring whole other modes of consciousness would have caught on without a shock like the assassination is hard to know.

And Bob Dylan might well have plugged in his guitar at the 1965 Newport Folk Festival, fusing folk music with rock and outraging the faithful at his sacrilege. One wonders, though, whether he would have energized young people as he did if they hadn't been shocked into a feeling of angry inwardness. Aggressive subjectivity has always been a strain of American culture, of course, but Dylan was the one who made it almost disreputable for a performer simply to sing a song. Ever since, the singer has generally been expected to write the song and to be making a statement. The limitations of Dylan's singing voice only under-scored his sincerity and made the personal statement more powerful. It was a prophetic stance, one that was most effective in troubling times.

Even the Vietnam War was under way when Kennedy was alive, though some argue that he might not have pursued it as his successor, Lyndon B. Johnson, did. I would argue otherwise. America's involve-ment grew from the same sort of assumptions of technocratic omnipo-tence that Conant applied to the high schools, and most of those making the policy were Kennedy-appointed fine-tuners. The United States seemed intent on patrolling the world as a high school principal walks a corridor—looking for trouble, bristling at perceived affronts to its author-ity. After a while, people, many of them young, started asking, "What's the point?"

While the Kennedy assassination signaled that unimaginable things could happen suddenly, the Vietnam War gave many of the things that did happen meanings they would not have had otherwise. Kennedy's killing had made the political personal. Vietnam made the personal politi-cal. Long hair, for example, became more than a fashion. It was a state-ment of defiance against militarized minds. And boys' long hair, which made so many adults apoplectic, stood as a protest against machismo, though many young women found out it didn't prevent the same young men from being male chauvinists.

When adults claimed not to be able to tell the boys from the girls, the long-haired young could only marvel at the limited perception of elders whose idea of masculinity was a haircut. And the slogan "Make love, not war" asserted that even casual sex could be a Statement.

The thing that made the Vietnam War meaningful, of course, was the possibility that you or your friends might actually have to go, fight, and perhaps die there. As during the Korean War, military service during the war could be delayed, simply by staying in high school or college. This was in keeping with the national policy of encouraging extended education.

Middle-class students went to college, which was what their parents expected anyway. Still, there seemed little margin for error. Take a se-mester to "find yourself," and you could find yourself in Vietnam, though even that danger was more theoretical than real. At least until they received their bachelor's degrees, most middle-class students were relatively safe.

It was different for those whose poor schooling or lack of money made college impossible. They went in large numbers at the age of eighteen. The draft frightened college students into not flunking out. And the war provided students with a claim to righteousness and an emotional intensity born of a sense of risk. But the young urban black men and rural and working-class white men who did not go to college were the ones who paid the price.

Obviously, it's not worth waging a war—especially one that increased social division—simply for the purpose of making the lives of young people in school feel more meaningful. There have to be better ways to let young people take meaningful chances and confront the fundamental issues of their lives. Still, there's little doubt that the conflicts and dangers of the 1960s brought into many young people's lives a sense that their actions—not at some distant time, but right then—mattered.

From 1964 and the emergence of the free speech movement at Berkeley, the media focused far more on colleges and universities—many of whose students were, of course, teenagers—than on the high schools. Heavy coverage of campus protest enabled disaffected young people all over the country and the world to know exactly what gestures, costumes, and attitudes were driving adults nuts at any given moment.

One thing that's important to keep in mind when making such broad-brushed accounts of a difficult time is that most students, both at high schools and colleges, were either spectators or minor participants. They might have bought the music, acquired some hippie threads for weekend use, smoked some joints, and availed themselves of a more permissive sexual atmosphere. Yet, even 1960s teens, who might have been expected to exaggerate their discontent when speaking to researchers, declared themselves to be believers in hard work, admiring of their parents, and hoping for a prosperous adulthood. They were not inclined to change the world, though quite a few declared that they were impatient to be done with high school.

Because high schools were, in general, more repressive environments than the college campuses, the struggles of young people in them often appeared to be more petty. There were only a handful of incidents whose scale was large enough to win widespread attention. One such was the student "blowout," walkouts that closed six Los Angeles high schools in 1967. Because five of these had predominantly Mexican-American student populations, and the other was predominantly black, it was seen more as a matter of racial tensions than youth discontent. In fact, the students' demands, the first two of which were free speech and free press, were similar to those aired less publicly at high schools in cities, towns, and suburbs throughout the country.

Typically, these were perceived not as fights over such grand abstractions, but as tests of how far students could go to make trouble, and how aggressively schools tried to make them adhere to their standards. Male students were suspended for sideburns that were too long, and then a year or two later for shoulder-length hair. High school newspapers were suppressed for expressing controversial opinions, which often meant opinions at odds with those of the principal or which might catch the attention of the school board. Student councils, which had been accustomed to planning dances and pep rallies, were increasingly challenged to become forums where questions about curriculum and school regulations could be debated.

In fact, such issues are not petty at all. Since at least the mid-nineteenth century, advocates of high schools had argued that they would teach citizenship. Controversies over dress, political expression, and school newspapers directly addressed the issue of when citizenship starts. They also raised the question of whether citizenship can be learned in an environment in which only a very narrow range of political and personal expression is allowed.

The most important legal precedent established by the student movements of the era was achieved not by protesters at any of the great universities but by high school students in Des Moines. In *Tinker v. Des Moines Independent School System* (1969), the U.S. Supreme Court, hearing an appeal of a student's plea to be allowed to wear a black armband in protest of the Vietnam War, declared that high school students are "persons" who "did not shed their constitutional rights to freedom of speech or expression at the schoolhouse door." The court added that student communication, especially of unpopular views, could lead to precisely the kind of disturbances the school district feared, but noted, "our history says that it is this sort of hazardous freedom—this kind of openness—that is the basis of our national strength and of the independence and vigor of Americans who grow up and live in this relatively permissive and often disputatious society."

This decision followed a 1967 Supreme Court decision requiring that those being tried in juvenile courts have the same right to counsel and other elements of due process as other citizens. This decision had some unforeseen consequences, as it opened the door for states to try young people as adults when it suited them. At the time, however, it was part of a trend both in Washington and in the states to expand the rights of the young.

This movement reached its climax in 1971, when the Twenty-Sixth Amendment to the U.S. Constitution was ratified, lowering the voting age to eighteen. While a handful of other states had already done so in the wake of World War II, this amendment set a national policy to define the threshold for political maturity at the same age as that of military service. And when several states lowered their drinking age to eighteen, it appeared that the nation was moving toward establishing that age as the effective age of maturity. (That trend was reversed during the 1980s.)

The *Tinker* decision had a particularly strong impact in the schools of large cities, where administrators were forced to tolerate the distribution of newspapers and other publications not sponsored or censored by the school. Radical high school students even established a service to exchange articles among such publications. High school principals had previously been as suspicious of "outside agitators" as segregationist southern governors were, but they saw little choice but to tolerate pamphleteering. One fairly typical tract, distributed by the High School Students Union in New York, complained, "The main thing that's taught to us in school is how to be good niggers, obey rules, dress in our uniforms, play the game, and NO, Don't be UPPITY." It went on to castigate student government as a sham and argue that ID cards and lavatory passes were instruments of oppression.

A subtler result of student activism in the schools was that it became easier for some educators to win the right to experiment. Some broke away from the established schools to establish free schools and other institutions that reflected their values. (Segregationist parents were doing much the same thing in the South by establishing private academies.) "Alternative" schools were established within large school systems, sometimes for pragmatic reasons. Philadelphia's Parkway "school without walls," which still exists, combined the desires to mobilize the cultural resources of an entire city, to free students to take greater responsibility for their own schooling, and to avoid the cost of erecting a new high school building. Less drastically, many high schools began to experiment with offering more electives and organizing existing classes around topics they hoped teenagers would find "relevant."

In 1973 the American century ran out of gas. Our inability to win in Vietnam had certainly not burnished the national sense of omnipotence. But the Arab oil embargo was probably the biggest affront to our pride since Japan bombed Pearl Harbor. Besides, the ongoing Watergate investigations, which drove President Richard Nixon out of office the next year, pointed to some problems of democracy that weren't taught in high school.

For teenagers born in the second decade of the baby boom, the world looked rather different from that of those born in the first. Their older brothers and sisters had come of age in an age in which anything seemed possible, money was available, and the future was limitless. The later baby boomers, those born in the late 1950s and early 1960s, came of age in the first period since the Depression in which scarcity was an issue. Progress was no longer assured, and, as the interstellar rattletraps of the 1977 film *Star Wars* suggested, even the Space Age was starting to rust.

The immediate scarcity of gasoline, which was causing long lines at the pump, was bad enough. But the consciousness of scarcity soon took on cosmic proportions. The entire world was "a population bomb." Even if nuclear war could be avoided, some said, overpopulation and the depletion of the earth's resources pointed toward a horrible war of all against all in competition for the earth's waning bounty. Economic growth was no longer a guilt-free solution to providing social justice. Teenagers were not merely being asked to wait a long time for their futures, but the quality of those futures were very much in doubt.

Sociologists who studied high schools in the 1970s found several lasting effects of the 1960s. One was that schools were far more willing to tolerate fashions of dress and grooming than they had been a decade earlier. Students were assuming a role they would play in the school—freak, politico, greaser, punk, cheerleader, "brain"—and carefully dressing and acting the part. Sociologists had once had to do extensive interviews to identify the groups in the school. Now they each had a look and a territory, though they worried about their lack of authenticity. "You have to change yourself if you want to have friends," a self-described female freak student at a wealthy suburban high school named "Sarah" told researcher Ralph W. Larkin in 1976. "It's not a very idealistic thing, like you should be, you are supposed to be yourself, you're supposed to be an individual. There are few people who can dare do that."

During the period of educational reform that followed the traumas of the 1960s, many high schools relaxed their systems of student control, as hall passes and required study halls gave way to "open campus" or "commons" arrangements. Students were allowed to spend some of the school day hanging out with one another, perhaps even smoking in designated areas. In most such schools, groups of students sharing particular interests established their own turf in the school building or grounds. Such territoriality is, perhaps, common animal behavior, but it belied the rhetoric of school spirit and social unity that had long been part of the ethos of high school.

The loosening of arbitrary rules and tight social controls within the schools was intended as a humanizing gesture (to teachers as well as students), and also a recognition of high school students' growing maturity. But not all experienced it that way. As teachers and administrators became less intrusive in students' lives, they also came to seem less concerned. Several different researchers who interviewed students in high schools from the mid-1970s onward found that one of the students' chief complaints was that nobody cared about them. They saw all the adults in the school—again with the frequent exception of athletic coaches— as detached and following their own agendas, while students were allowed to drift. The schools had surrendered their role as substitute parents at a time when parents themselves were increasing their working hours and reducing the amount of time they spent with their families.

The result was a strengthening of precisely what both schools and parents feared: the peer group culture. By requiring large numbers of young people to be physically present, while leaving part of the day unscheduled, some schools created an environment for dealing marijuana and other illegal substances, organizing gang activities, smoking cigarettes, and other undesirable behavior.

While some believe that this reformist moment in the 1970s was when American education took a decisive turn for the worse, others regret that an opportunity to rethink secondary schooling was lost. It is, however, a mistake to overstate both the changes that came in the schools and the influence they had. Although schools purport to prepare students for a changing world, they are themselves notoriously resistant to change. Once student protests abated and the media moved on to new crises, the schools reverted to being somewhat more nervous, slightly less complacent versions of what they had been when the trouble started. But in the two areas that most affect the school—the economy and the family—conditions were changing radically.

The use of drugs, especially marijuana, was another lasting effect of the 1960s. Unlike their predecessors, however, the 1970s students didn't talk so much about expanding their minds as dulling their senses, so they could make it through yet another boring day. Use of marijuana and other illegal drugs, along with the use of alcohol, was generally on the rise among teenagers throughout the 1970s.

Parents' suspicions that their children were using marijuana were helping to widen what, real or not, was perceived as a generation gap. "Sarah" told Larkin that she could use marijuana and fool her mother by acting vaguely angry, but when she was happy and hadn't used drugs, her mother invariably accused her of being high.

Some have argued that the hiding of certain activities—particularly the use of intoxicants and sexual activity—from elders is a universal phenomenon of adolescence. It may be more accurate to say that it's true of societies, like those African groups discussed earlier, that practice very strict age group segregation. Elders are not willing to acknowledge "adult" behavior by the young because they are not willing to acknowledge young people's maturity. Thus, they avert their eyes, fear the worst, hope for the best, and maintain an atmosphere of mistrust.

In recent years, the noncommunication and suspicion between generations have become so extreme that some families have instituted a practice of random drug testing of their children. This is a historic reversal. While schools, reformatories, and police once derived their authority from the family, now parents are trying to find power in mimicking the impersonality and harshness of public bureaucracies.

Meanwhile, young people are doing relatively little to hide their sexual activity, though parents are inclined to avert their eyes anyway. Comprehensive studies of contemporary teenagers' sexual activity began to appear during the 1970s. (The Kinsey and other data cited in earlier chapters were based on memories that were often decades old.) What they said was something that parents knew but didn't really want to know: that the majority of women have sex during their teens. (So do males, of course, but women are more often worried about—a phenomenon that the feminist movement has done little to change.)

"Going steady," with its implications of trial marriage, had begun to disappear during the 1960s. By the 1970s, young people increasingly socialized in groups, and they saw sex not so much as a prelude to marriage but as a form of personal exploration and intimate communication. They still got hurt when a boyfriend or girlfriend left them for someone else, but they were too realistic to think that a relationship at fifteen was going to last until the couple was ready to marry ten years hence.

Because virginity was no longer highly prized, and other young people, adults, and the news media seemed to be talking about sex constantly, there was a pressure to get on with it. When Offer did a new survey of high school students toward the end of the 1970s, the great majority of their attitudes were similar to those of their counterparts in the early 1960s. One exception was that a far higher percentage of 1970s young people worried that they were "way behind" their peers when it came to sex. This admission didn't correlate with anxiety over being somehow abnormal, but it does indicate some sense of inadequacy about not being sexually active in a society where everyone else seems to be doing it incessantly.

In fact, teenage sexuality was on the increase: 36 percent of fifteen- to nineteen-year-old women had had intercourse in 1975. The figure rose to 47 percent in 1982, 53 percent in 1988, 55 percent in 1990, then fell to 50 percent in 1995, and 47.7 percent in 1997. During the 1980s, even the compiling of such data became politically controversial; opponents believed that simply asking the questions would give teenagers ideas. The same unlikely assumption underlies efforts to prevent schools from teaching about contraception and avoiding sexually transmitted diseases. It's more likely that the combination of sex-saturated commercialism and the pressures to feel accepted and grown up are to blame. As Andrea N. Jones, then a teenager, wrote in a 1994 article in *Youth Outlook,* "If adults use sex to sell toiletries, why shouldn't kids use it to sell themselves?"

The later baby boomers were, of course, influenced by the behavior of the first generation of baby boomers as they proceeded into their twenties.

Those who were born after the first six or seven years of the baby boom experienced most of the disadvantages of crowding and anonymity that earlier boomers had felt without the compensating sense of being a pioneer. What was worse, they also suffered to some extent from a rather mysterious cultural phenomenon—a backlash against the young.

An early symptom of this phenomenon was the 1968 film *Rosemary's Baby,* in which Mia Farrow discovered that she was about to have a demon child. Throughout the 1970s, such films as *The Exorcist, The Omen,* and their sequels and knockoffs, such as *Children of the Corn,* cast children as the carriers of evil. These movies coincided with a sudden loss of momentum for education reform, and waning political support for schools, even in suburban areas where they had once been sacrosanct. School board voters wanted to get the schools back to basics, and they prized order and economy over experimentation. Birthrates, which, already on the way down, fell more sharply, even though boomers were reaching the age at which many of their parents had given birth to them.

One could argue that this antichild backlash reflected a certain weariness with the baby boom, who despite their privileged upbringing had proven to be, if not demonic, then troublesome. The problem with this theory, though, is that early boomers seem to have been leaders of the backlash. They were the ones going to these movies, and they were the ones not having children. They had convinced themselves not to trust adults, and having children is certainly a way to remind yourself you're not young anymore.

For parents of some teenagers, these possessed children from the movies were a horrifying prophecy, as they came to fear "losing" their children to religious and political organizations that required total devotion and alienation from family and others outside the community. These so-called cults—which ranged from Christian fundamentalist revivalism to the exoticism of the saffron-robed, shaved-headed Hare Krishnas, the Space Age visions of Scientology, and the political paranoia of the followers of Lyndon LaRouche—were able to win the total commitment of their followers. They demanded a discipline not required at home or school, but perhaps most important, they gave young people key roles in struggles of cosmic import. While the cults claimed the allegiance of only a small percentage of the young, the numbers were still large, and some groups made themselves very visible in public places. Even when there was no costume, you could pick them out in a crowd of young people because they seemed to show no anxiety whatsoever.

Like the the pounding, rhythmic, heavy-metal music that first became popular during the 1970s, the youth cults offered the promise that you could lose yourself and be yourself at the same time. They spoke to the suspicion of some young people that contemporary life was unreal and meaningless. As at the time of the Great Awakening in the eighteenth century and the second Great Awakening in the nineteenth century, spiritual longing took the young in some directions their elders couldn't understand.

FOURTEEN
Goths in Tomorrowland

I feel stupid and contagious.

—KURT COBAIN, "Smells Like Teen Spirit" (1991)

In the summer of 1997, the security forces at Disneyland and the police in surrounding Anaheim, California, announced a "zero tolerance" policy to fend off a new threat.

Hordes of pale, mascaraed goths—one of the many tribes of teendom—were invading. It was an odd onslaught. Unlike their barbarian namesakes, they weren't storming the gates of the walled Magic Kingdom. They had yearly passes, purchased for $99 apiece. Many of them had not even been goths when their parents dropped them off at the edge of the parking lot. Rather, they changed into their black sometimes gender-bending garments, applied their white makeup accented with black eyeliner and gray blush-on. The punkier among them accessorized with safety pins and other aggressively ugly, uncomfortable-looking pierceables. And most important of all, they reminded themselves to look really glum. Once inside, they headed for Tomorrowland, Disneyland's most unsettled neighborhood, and hogged all the benches.

It was a sacrilege. Disneyland, said those who wrote letters to the editor, is supposed to be "the happiest place on earth," and these young people with their long faces clearly didn't belong. The presence of sullen clusters of costumed teens showed, some argued, that Disney had given up its commitment to family values. It was no longer possible to feel safe in Disneyland, came the complaints, and that was about the last safe place left.

Actually, the safety of Disneyland was part of the attraction for the goth teens. They told reporters that their parents bought them season passes because the theme park's tight security would assure that nothing bad would happen to them. In the vast sprawl of Orange County, California, there are very few safe places where teens are welcome, and Disneyland has always been one of them.

Those who complained spoke of the goths as if they were some sort of an alien force, not just white suburban California teenagers. Only a few years earlier, they had been kids who were delighted to go with their parents to meet Mickey. And only a few years from now, they will be young adults—teaching our children, cleaning our teeth, installing our cable television. But now they insist on gloom. And the adult world could not find a place for them—even in Tomorrowland.

Unlike Minnesota's Mall of America—which became a battleground for gang warfare transplanted from Minneapolis and which eventually barred unescorted teenagers from visiting at night—the perceived threat to Disneyland was handled in a low-key way. Teenagers were arrested for even the tiniest infractions outside the park and forced by security guards to follow Disneyland's quite restrictive rules of decorum within the park. After all, the theme park's administrators had an option not available to government; they could revoke the yearly passes. While Disneyland doesn't enforce a dress code for its visitors, it can keep a tight rein on their behavior.

Yet, despite its lack of drama, I think the situation is significant because it vividly raises many of the issues that haunt teenagers' lives at the end of the twentieth century. It is about the alienation of teenagers from adult society, and equally about the alienation of that society from its teenagers. The mere presence of teenagers threatens us.

It is also a story about space. How, in an environment devoid of civic spaces, do we expect people to learn how to behave as members of a community? And it is about the future. Is a meaningful tomorrow so far away that young people can find nothing better to do than engage in faux-morbid posturing? (Even Disney's theme parks are losing track of the future; they are converting their Tomorrowlands into nostalgic explorations of how people used to think about the future a century and more ago.)

And even its resolution—a stance of uneasy tolerance backed by coercion and force—seems symptomatic of the way Americans deal with young people now.

Inevitably, a lack of perspective bedevils efforts to recount the recent past, but the problem is more than that. The last quarter of the twentieth century has, in a sense, been about fragmentation. Identity politics has led to a sharpening of distinctions among the groups in the society, and a suspicion of apparent majorities. Postmodern literary theory warns us to mistrust narratives. Even advertising and television, which once united the country in a common belief in consumption, now sell to a welter of micromarkets. Thus are we left without either a common myth, or even the virtual common ground of "The Ed Sullivan Show."

It seems crude now to speak of teenagers and think of the white middle-class, heterosexual young people that the word "teenager" was originally coined to describe. The "echo" generation of teenagers, whose first members are now entering high school, is about 67 percent non-Hispanic white, 15 percent black, 14 percent Hispanic, and 5 percent Asian or American Indian. The proportion of Hispanic teens will grow each year, and the Census Bureau also reports significantly greater numbers of mixed-race teens and adoptees who are racially different from their parents.

Even the word "Hispanic" is a catch-all that conceals an enormous range of cultural difference between Mexicans, Cubans, Puerto Ricans, Dominicans, and other groups whose immigration to the United States has increased tremendously during the last quarter century. Urban school systems routinely enroll student populations that speak dozens of different languages at home.

Differences among youth do not simply involve differences of culture, race, income, and class—potent as these are. We now acknowledge differences in sexual orientation among young people. Today's students are also tagged with bureaucratic or medical assessments of their abilities and disabilities that also become part of their identities.

There are so many differences among the students at a high school in Brooklyn, Los Angeles, or suburban Montgomery County, Maryland, that one wonders whether the word "teenager" is sufficient to encompass them all. Indeed, the terms "adolescent" and "teenager" have always had a middle-class bias. In the past, though, working-class youths in their teens were already working and part of a separate culture. Now that the work of the working class has disappeared, their children have little choice but to be teenagers. But they are inevitably different from those of the postwar and baby boomer eras because they are growing up in a more heterogeneous and contentious society.

What follows, then, is not a single unified narrative but, rather, a sort of jigsaw puzzle. Many pieces fit together nicely. Others seem to be missing. It's easier to solve such a puzzle if you know what picture is going to emerge, but if I were confident of that, I wouldn't be putting you, or myself, to such trouble.

These discussions do have an underlying theme: the difficulty of forging the sort of meaningful identity that Erik Erikson described at midcentury. But if we look for a picture of the late-twentieth-century teenager in these fragments, we won't find it. That's because we're expecting to find something that isn't there.

The goths who invaded Tomorrowland are examples of another kind of diversity—or perhaps pseudo-diversity—that has emerged gaudily during the last two decades. These are the tribes of youth. The typical suburban high school is occupied by groups of teens who express themselves though music, dress, tattoos and piercing, obsessive hobbies, consumption patterns, extracurricular activities, drug habits, and sex practices. These tribes hang out in different parts of the school, go to different parts of town. Once it was possible to speak of a youth culture, but now there is a range of youth subcultures, and clans, coteries, and cliques within those.

In 1996 a high school student asked fellow readers of an Internet bulletin board what groups were found in their high schools. Nearly every school reported the presence of "skaters," "geeks," "jocks," "sluts," "freaks," "druggies," "nerds," and those with "other-colored hair," presumably third-generation punks. There were also, some students reported, "pager people," "snobs," "band geeks," "drama club types" (or "drama queens"), "soccer players" (who aren't counted as jocks, the informant noted), "Satanists," "Jesus freaks," "industrial preps," "techno-goths," and "computer dweebs." Several took note of racial and class segregation, listing "blacks," "Latinos," "white trash," and "wannabe blacks." There were "preppies," who, as one writer, possibly a preppie herself, noted, "dress like the snobs but aren't as snobbish." "Don't forget about the druggie preps," another writer fired back.

This clearly wasn't an exhaustive list. Terms vary from school to school and fashions vary from moment to moment. New technologies emerge, in-line skates or electronic pagers for instance, and they immediately generate their own dress, style, language, and culture.

The connotations of the technologies can change very quickly. Only a few years ago, pagers were associated mostly with drug dealers, but now they've entered the mainstream. Pagers became respectable once busy mothers realized that they could use them to get messages to their peripatetic offspring. Young pager users have developed elaborate codes for flirtation, endearment, assignation, and insults. They know that if 90210 comes up on their pager, someone's calling them a snob, and if it's 1776, they're revolting, while if it's 07734, they should turn the pager upside down and read "hELLO."

Most of the youth tribes have roots that go back twenty years or more, though most are more visible and elaborate than they once were. Many of these tribes are defined by the music they like, and young people devote a lot of energy to distinguishing the true exemplars of heavy metal, techno, alternative, or hip-hop from the mere poseurs. Hybrid and evolutionary versions of these cultures, such as speed metal, thrash, or gangsta rap make things far more confusing.

One thing that many of these subcultures have in common is what has come to be known as modern primitivism. This includes tattooing, the piercing of body parts, and physically expressive and dangerous rituals, such as the mosh pits that are part of many rock concerts. Young people use piercing and tattoos to assert their maturity and sovereignty over their bodies.

"Can this be child abuse?" Sally Dietrich, a suburban Washington mother, asked the police when her thirteen-year-old son appeared with a bulldog tattooed on his chest. "I said, 'What about destruction of property?' He's my kid." Her son was, very likely, trying to signal otherwise. Nevertheless, Dietrich mounted a successful campaign to bar tattooing without permission in the state of Maryland, one of many such restrictions passed during the 1990s.

It may be a mistake to confuse visible assertions of sexual power with the fact of it. For example, heavy-metal concerts and mosh pits are notoriously male-dominated affairs. And the joke of MTV's "Beavis and Butt-head" is that these two purported metalheads don't have a clue about how to relate to the opposite sex. Those whose costumes indicate that they have less to prove are just as likely to be sexually active.

In fact, visitors to Disneyland probably don't need to be too worried about the goths, a tribe, which like many of the youth culture groups, has its roots in English aestheticism. As some goths freely admit, they're pretentious, and their morbid attitudes are as much a part of the dress-up games as the black clothes themselves.

The goth pose provides a convenient cover. For some males, it gives an opportunity to try out an androgynous look. The costumes, which emphasize the face and make the body disappear, may also provide an escape for young women and men who fear that they're overweight or not fit. Black clothes are slimming, and darkness even more so. "Until I got in with goths, I hadn't met other people who are depressed like I am and that I could really talk to," said one young woman on an Internet bulletin board. Another said being a goth allowed her relaxation from life as a straight-A student and a perfect daughter.

Although young people recognize an immense number of distinctions among the tribes and clans of youth culture and are contemptuous of those they regard as bogus, most adults cannot tell them apart. They confuse thrashers with metalheads and goths because they all wear black. Then they assume that they're all taking drugs and worshipping Satan.

The adult gaze is powerful. It classes them all as teenagers, whether they like it or not. The body alterations that young people use to assert that they are no longer children successfully frighten grown-ups, but they also convince them these weird creatures are well short of being adults. The ring through the lip or the nipple merely seems to demonstrate that they are not ready for adult responsibility. What they provoke is not respect but restrictions.

Tribes are about a yearning to belong to a group—or perhaps to escape into a disguise. They combine a certain gregariousness with what seems to be its opposite: a feeling of estrangement. The imagery of being alone in the world is not quite so gaudy as that of modern primitivism. yet it pervades contemporary youth culture.

While youthful expression of the 1920s, 1940s, 1950s, or 1960s often took the form of wild dancing, more recently it has been about solitary posing. This phenomenon is reflected, and perhaps encouraged, by MTV, which went on the air in 1981. In contrast with the rudimentary format of "American Bandstand," in which the viewer seemed simply to be looking in on young people having fun dancing with one another, MTV videos tend to be more about brooding than participation. They are highly subjective, like dreams or psychodramas. They connect the viewer with a feeling, rather than with other people.

And while the writhing, leaping, and ecstatic movement of the mosh pit seems to be an extreme form of "American Bandstand"-style participation, it embodies a rather scary kind of community. One's own motions have little relationship to those of others. And there's substantial risk of injury. The society implied by the dance is not harmonious and made up of couples. Rather, it is violent and composed of isolated individuals who are, nevertheless, both seeking and repulsing contact with others. If this sounds like a vision of American society as a whole, that's not surprising. Figuring out what things are really like is one of the tasks of youth. Then they frighten their elders by acting it out.

When a multinational company that sells to the young asked marketing psychologist Stan Gross to study teenagers around the country, he concluded after hundreds of interviews and exercises that the majority of young people embraced an extreme if inchoate individualism. Most believe that just about every institution they come in contact with is stupid. When asked to choose an ideal image for themselves, the majority selected a picture that depicted what might be described as confident alienation. The figure sits, comfortably apart from everything, his eyes gazing out of the image at something unknown and distant.

Such studies are done, of course, not to reform the young but to sell to them. And the collective impact of such knowledge of the young has been the proliferation of advertising that encourages young people not to believe anything—even advertising—and to express their superiority by purchasing the product that's willing to admit its own spuriousness.

The distance between spontaneous expression and large-scale commercial exploitation has never been shorter. Creators of youth fashion, such as Nike, go so far as to send scouts to the ghetto to take pictures of what young people are wearing on the streets and writing on the walls. Nike seeks to reflect the latest sensibilities, both in its products and its advertising. The company feeds the imagery right back to those who created it, offering them something they cannot afford as a way of affirming themselves.

One result of this quick feedback is that visual symbols become detached from their traditional associations and become attached to something else. Rappers, having made droopy pants stylish in the suburbs, began to wear preppie sportswear, and brand names like Tommy Hilfiger and Nautica became badges of both WASP and hip-hop sensibilities. Thus, even when the fashions don't change, their meaning does. Such unexpected shifts in the meaning of material goods cannot be entirely manipulated by adults. But marketers have learned that they must be vigilant in order to profit from the changes when they come.

More overtly than in the past, many of today's young are looking for extreme forms of expression. This quest is just as apparent in sports, for example, as in rock culture. The 1996 Atlanta Olympics began with an exhibition of extreme cycling and extreme skating. These and other extreme sports, categorized collectively as "X-Games," have become a cable television fixture because they draw teenage males, an otherwise elusive audience. "Extreme" was one of the catchwords of the 1990s, and it became, by 1996, the most common word in newly registered trade names, attached either to products aimed at youth or which sought to embody youthfulness.

Young people are caught in a paradox. They drive themselves to extremes to create space in which to be themselves. Yet, the commercial machine they think they're escaping is always on their back, ready to sell them something new.

Still, even as many teenagers have been struggling to adopt ever more outré costumes and attitudes that have stirred adult fears, their relatively small numbers and their political quiescence have made them easy to ignore. For most of two decades, communities have been able to get by without expanding their high schools or worrying much about their schooling at all. Rather than protest, young people have simply disengaged. Those over eighteen can now vote, but they don't do so in large numbers, and when they do, they tend to vote like their parents.

There are, in fact, political attitudes buried in the deeply pessimistic symbolism that several of the tribes of youth affect. They combine extreme libertarianism with a belief that nearly anything that could be done about whatever is wrong now will only make it worse. Today's teenagers are literally children of Ronald Reagan's America, and despite a moodiness that contrasts with the former President's endearing sunniness, they have absorbed many of its political assumptions as well.

More than six out of ten contemporary teenagers have grown up without a parent at home during the day. That's a far cry from the baby boomers' experience. That doesn't necessarily mean that these young people were ignored or neglected by their parents. It does mean that, in many cases, their ties to their parents are looser, provoking worries that they might be led astray by peers. In contrast with their libertarian children, parents have increasingly looked to government to enact and enforce curfews and to place other restrictions on the young. They expect government to discipline their children as parents themselves once did.

For most families, long work hours and parental absence from the home are not an option but a necessity—at least if the family is to maintain a high material standard. The work is, in fact, often justified in terms of providing young people with the best opportunities for schooling and later life. Still, there must be some connection between the sense of disengagement that adults bemoan in contemporary teenagers and young people's complaints that their elders don't simply misunderstand them but really can't be bothered with them.

There's no doubt that peers are influential in key areas of young people's lives, no matter what parents do. In a society like ours, where change is rapid and teenagers spend most of their time with others exactly their age, the young have more authority than adults have. Still, there is evidence that if parents to take a lively, though not defensive, interest in their children's lives, their teens are less likely to commit crimes, use drugs, or become pregnant prematurely. For example, teenagers who have dinner with their families most nights are far less likely to get into trouble than those who do not. Yet, the pressures on both parents and teenagers are in the other direction: toward immersion in work to support the overhead, not the essence, of family life.

It is widely believed that there has been a massive increase in antisocial behavior, risk-taking, and sexual activity by teenagers in recent years. This is not accurate. A real upswing in such behavior occurred during the 1970s. For most kinds of problems, it reached a plateau in the early 1980s and remained quite high by historical standards into the 1990s, when there were signs of a reduction in crime, alcohol abuse, premarital pregnancy, and other problems.

Key exceptions to this pattern were arrests for serious crimes, which increased 70 percent during between 1985 and 1994, and homicides, which more than doubled during the same period. Arrest figures are always influenced by the number and aggressiveness of police available at a given moment, and the rise in crime may not be as dramatic as the figures suggest. But the figure for homicides does indicate that violent crimes committed by teenagers have, in fact, become more violent.

Although most of the problems of young people have been around for a long time and don't seem to be getting much worse, the situation is far from ideal. Despite the democratic ideals that underlie the educational system that creates and defines teenagers, the United States seems more willing than many other countries to let large numbers of young people drift into wasted lives of idleness and crime. We also support the largest incarcerated population in the world.

Thus, even as the attention adults pay to individual teenagers has apparently declined, the degree to which adults fear them as a group has unquestionably increased. The result has been, among other things, the enactment of laws that deny them, as minors, freedom to move, gather, and express themselves, and of other laws that require states to prosecute them as adults for a wide variety of serious crimes.

The first group of laws are based on the hope of perfectibility I discussed in the first chapter. The teen years, the argument goes, are the time when people are susceptible to falling into bad habits and addictions. Although problems of alcohol and drug addiction, drunken driving, and out-of-wedlock pregnancy are actually worse among those in their twenties than among teenagers, teenagers are targeted because they may be more reformable, and because as minors, they have fewer legal rights.

This approach has rarely been successful. "Do as I say, not as I do" is an admonition that has never worked. And the message that a particular activity is for grown-ups only, far from suppressing the behavior, actually encourages it. Aggressive efforts against teenage smoking during the 1990s have coincided with the first increase of such smoking in decades.

Prohibition encourages resourcefulness as young people learn how to get around the law. When, in 1997, a zero tolerance enforcement campaign on selling tobacco to minors—using teenage decoys to attempt to buy cigarettes—was tested in Massachusetts, it produced arrests but absolutely no decrease in the amount or ease of teenage smoking. Another study of drug use, primarily marijuana, among teenagers indicated that high school—the most restrictive environment most teenagers experience—is also the most common place where drugs are purchased.

Adults want people in their teens to be perfect, but young people have no such interest. They'd rather be grown up.

The second set of laws reflect an opposite assumption from the first. They assume that young people *are* grown up, and that they should face the same judicial standards and punishment for serious crimes as any adult. This approach is politically attractive, and it appears not to be as arbitrary and capricious as laws that are restricted to juveniles. Nevertheless, nobody has been able to demonstrate that such laws have a deterrent effect. Many prosecutors oppose such laws because they have found that juries are reluctant to mete out harsh punishment to the young. Nevertheless, many large cities have significant numbers of incarcerated young, many of whom could not afford to pay for lawyers to defend them.

The most successful programs for curbing juvenile crime—most notably one that virtually eradicated teenage homicides in the city of Boston—revive the premise of the juvenile justice system as it was first conceived a century ago. In Boston probation officers frequently ride with policemen on their nightly patrols. They know the young people they are supervising personally, and they can coach them, negotiate with them, coerce them, or cite them for violating the terms of their probation and have them jailed. The probation officers' attention, while not always welcome, is nevertheless personal. Young people, like most of the rest of us, tend not to respond well to bureaucratic imperatives; they do respond to other people.

Such initiatives were accompanied by the opening of youth centers, strong participation by churches and other community groups, crackdowns on gang activity, and specially assigned police officers lecturing in schools and leading field trips. All were designed, according to one of the leaders of the effort, "to strip away the anonymity" of youthful offenders.

There are cities where this sort of approach has gone overboard and produced an atmosphere where even those who have never misbehaved feel threatened. The juvenile court approach has been coercive since its inception, and based on the premise that the civil rights of minors are very limited.

Moreover, such approaches do not deal with the reality that large numbers of young people, especially African Americans and Hispanics, are growing up in neighborhoods that are essentially without any legal economic activity. About 43 percent of black children and 41 percent of Hispanic children lived in poverty in 1994, and the percentage of all children in poverty has risen by about 50 percent since 1970. Yet, those who grow up in these impoverished economic dead zones—the same places Nike scouts for style—are subject to the same pressures for acquisition and display as the everyone else. Indeed, some rap performances consist of little more than a litany of brand names of luxury goods that can only be obtained by a successful rap artist, a star athlete, or a gangster. These are the only ladders of success that many young people can imagine.

For poor youths, it's very dangerous to be young. For young black males, homicide, not automobiles, is the leading cause of death, at a rate of 8.4 per 100,000. Urban poverty has always spawned violence, but now it's more likely to kill.

Something that is often forgotten amid fear of violent youth crime is that people in their teens, in all income classes, are victims of violence far more frequently than adults. Some of this violence comes at the hands of their peers, but much is the work of adults. In a representative national sample of ten- to sixteen-year-olds, one in four reported incidents of abuse during the previous year.

In most of the earlier chapters, we've seen how money and economic opportunity have shaped family life and the expectations parents have for their young. This remains true. And for families, the single most important economic fact of the last quarter century has been the decline of real, inflation-adjusted individual wages. This decline in wages has deeply affected the lives of young people.

Median annual income, adjusted for inflation, fell 10 percent for all male workers between 1975 and 1994. For a large number of families, this decline meant that the one-income family that was the norm when teenagers first came into existence was no longer economically possible. That's not the only reason so many women came into the workforce during that period, but it is a significant one. Because of women working, family income has continued to rise throughout the last two and a half decades, but family experience has certainly changed.

Moreover, while women have gained greater income parity with men, their median income has also declined. While the single working mother can console herself that she is being paid comparably to her male coworker, she nevertheless earns less to support her household than she would have two decades ago.

Despite superficial prosperity, families in middle- and lower-income ranges have been suffering from a two-decade-long squeeze, one that helps explain their inability to spend as much time with their children as middle-class norms have traditionally required.

What contemporary American families are attempting—providing a prolonged, protected period of childhood and youthful preparation for our offspring while both parents work outside the household—is novel. In the past, when both parents worked, their children did so as well. On farms, work and family life were essentially inseparable. And when schooling became the job of the young, it was usually supported by a nonworking mother maintaining a household that explicitly supported educational values. Today the only constant presence in the household is the television set, which instructs the children not on what they should become, but rather on what they should buy.

Our attempt to maintain an elaborate domestic life—with nobody home—is an experiment that seems doomed to fail. In practice, people are getting by. Children are growing up, as they always have, more or less okay. Yet, when young people appear to be less stable and more troubled than they seemed to be four decades ago, we ought to acknowledge that both adults and their children are living far more precarious lives.

The wage rate squeeze also plays a major role in another of the familiar stories about recent teenagers—the one about precocious pregnancy.

As I've noted earlier, the overall rate of teenage pregnancy was significantly higher during in the 1950s than it is today. The difference is that, in the 1950s, marriage either preceded or followed soon after conception. Now the great majority of teenage pregnancies involve women who aren't married.

In today's economy, however, marriage is rarely a "solution" to teenage pregnancy. You could accost the young man who's the father of your daughter's baby with the traditional shotgun, but even at gunpoint, he probably won't be able to find a job that will pay enough to support a new family. The wage rate decline that began in the mid-1970s affected young people diproportionately. Median incomes dropped 26 percent for teenage males between 1975 and 1994, and 36 percent for those in their early twenties.

Moreover, if the young man hasn't finished high school, many employers will refuse even to consider him. His prospects are bleak. One out of three noncompleters of high school has no income at all the following year. Even for high school graduates, unemployment of those under twenty tends to average around 18 percent for whites, and nearly twice that for blacks. While education doesn't necessarily prepare young people for jobs, it has become a virtual prerequisite for anyone who plans to raise a family.

In fact, your daughter might be making a rational decision in not burdening herself with a husband, even though her immediate income expectations are not as good as those of a young man her age. The point is not that fathers are unnecessary in a household; close to 60 percent of female-headed single-parent households are in poverty. But it is also true that men aren't the economic saviors they once were. They are, at best, only part of the solution. If there's a grandmother available to take care of the baby while the mother works or goes to school, a young woman can be just about as well off as if she had a husband.

The so-called epidemic of teenage pregnancies is part of a larger trend by American women in all ethnic and racial groups to bear children out of wedlock. Similar trends are happening in other countries where women are approaching economic parity with men. Childbearing has become increasingly dissociated from marriage.

As the historian Constance Nathanson has pointed out, the current tendency to analyze the impact of adolescent pregnancy in terms of the mother's loss of opportunity for education and future earning power represents a recent and radical change in our thinking. The idea that motherhood and the making of a household were central tasks of womanhood virtually defined middle-class values from the 1820s onward. Now the ill-timed child is seen primarily as an impediment to the young woman's own career aspirations.

And the 1996 welfare reforms, which were intended to force the mothers of even small infants into the paid workforce, suggest that, as a society, we have become blasé about America's babies.

Out-of-wedlock motherhood among teenagers hasn't merely increased. It has become far more visible. One important reason for that was the enactment in 1972 of the antidiscrimination legislation known as Title IX. Before Title IX, young women who became pregnant withdrew from high school or were routinely expelled. Motherhood was seen as completely incompatible with being a student. Moreover, the pregnant girl was believed to set a bad example for her fellow students. Shamed parents frequently sent their children to stay with relatives, or otherwise kept her from being visible to the community.

Title IX, which was intended to guarantee equal treatment of girls and boys in schools and colleges, changed all that. Young women could no longer be barred from publicly funded schools because they were pregnant. Some school systems set up separate programs for teenage mothers; others provided day care and have allowed the young women to continue their usual high school courses.

The issues involved in a teenager's pregnancy—whether to abort, allow adoption, or raise the child, and, if so, how—are no longer whispered or unspoken. They are present in the high school for all to see. The consequences of this new openness about teenage pregnancy are mixed. The young mother no longer faces the ostracism of society. But at the same time, she becomes a public issue—indeed, one of the recent villains of American politics.

This change of policy toward young women has helped to change the nature of high school. It was once viewed as a protective place, a place of preparation for future responsibilities. Young people who behaved as adults—either by working full-time or becoming parents—were excluded. That was all right as long as a high school diploma was not the minimum credential for making a living. Now that getting through school is virtually mandatory, the school must accommodate more kinds of students. Having babies and getting jobs are two things teenagers have always done. Now they do those things—and go to school too.

Title IX has also had another important consequence that affected far larger numbers of students. It brought about a surge in female participation in high school sports. Reluctantly at first, schools were forced to change the pattern of well-funded, well-coached athletic programs for boys, and token, stepchild programs for a marginal group of girls. Gradually, female sports have become more competitive, as assumptions about the girls' limited physical capacities have been jettisoned, along with rules that once tended to make such sports as women's basketball intrinsically dull.

The results have been staggering. In 1972, the year Title IX was enacted, about one high school girl in twenty-seven participated in team sports at her school. By 1997, that figure had risen to one girl in three. Athletics still aren't as desirable an activity for girls as for boys, but the stigma that previously accompanied women's sports has completely disappeared. Although physical education experts warn that the two thirds of young women who do not play team sports live more sedentary lives than did their counterparts of a quarter century ago, athleticism has become a desirable female attribute.

Perhaps as an indirect result of Title IX—and more directly, though the influence of such role models as Jane Fonda and Madonna—the ideal female body has become ever more difficult for young women to approximate. It is essential to have almost no fat, along with a high degree of muscle tone. Each historical moment generates its characteristic diseases, and by the early 1980s, the eating disorders anorexia nervosa and bulimia had become potentially life-threatening epidemics. The "germs" that caused them were, in large part, pictures in magazines.

For the first time in American memory, males too were feeling pressure to live up to idealized models. Feminists had complained that men looked upon women as nothing more than sex objects. An unanticipated consequence of the increasing equality of the sexes was that the male body became as eroticized and objectified as the female's. By the early 1990s, the image of Marky Mark in his briefs—including the parts that were concealed but definitely suggested—posed a challenge to young men everywhere. And few could measure up. Even as competitive sports were becoming a somewhat less important part of high school boys' lives, competitive bodies were becoming more important. Moreover, the ideal male body type was not, as it had traditionally been, a normal by-product of athletics, work, and physical activity. Rather, it was something to be consciously achieved through exercise intended to make the body conform to an aesthetic ideal.

It is, then, not surprising that one area in which the mental state of the normal teenagers Daniel Offer surveys has shown a recent turn for the worse is in boys' body image. Young males still, on the whole, feel very confident of and take pride in their bodies, but by not quite so overwhelming a margin as in the past. Young males may be moving toward the greater insecurity that females have expressed for three decades.

In many places, the shower after gym class—a high school ritual since World War I—disappeared, especially in the boys' locker rooms. One reason the young men give is fear that their gay classmates might be looking at them, which wasn't a problem back when it was generally assumed that homosexuals didn't exist in such "normal" settings as high school. The more basic problem is that they are aware that everybody's looking, something that female classmates have always known. (Showering after gym seems to be more prevalent now among females than males. Young women still fear that they will offend if they go unwashed, while young men worry less about such matters.)

One of the biggest changes in the experience of teenagers during the last two and a half decades has been the increasing role that paid work plays in their lives. A significant majority of teenagers have jobs, beginning in their sophomore year in high school, and the length of the hours they work has been increasing, especially for females, who had previously worked less than males. Work weeks of twenty hours and more are not uncommon, nor are shifts to midnight and beyond in the fast-food outlets and convenience stores where many young people work. If you add twenty hours a week to thirty at school, there is little time or energy left for homework, which some believe has become easier simply because teachers know their students can't handle it.

During the 1930s, a lack of work drove young people into the high schools—setting the stage for the emergence of the teenager as we know it. Teenagers were to be protected from work so that they could prepare themselves. And the labor market was to be protected from young workers, who would drive down wages and make it more difficult for adult workers to raise their families. In this regard, as in so many others, today's young people are no longer protected, nor are the wages of adults who might otherwise work in retailing and fast-food jobs where teenagers predominate.

Yet, unlike other behavior that deviates from what was once expected of teenagers, their increased workload provokes neither fear nor criticism. Americans, young and old, believe in work. More than 90 percent of teenagers place a high value on working hard and doing a good job. Parents tend to endorse work as a character-building exercise. They also believe that their children will appreciate a thing they want more if they are required to work for it.

Yet, the percentage of teenagers' earnings that are contributed to the needs of the family as a whole is under 10 percent on average, and zero for most. The percentage saved for college is larger, but still modest. Paying for automobiles, clothes, and entertainment are much more common uses for teenagers' earnings. Teenagers from poorer families are actually less likely to work than those in middle-class families, in part because jobs for teenagers are more prevalent in suburban areas.

The issue of whether teenage work is a good thing is controversial. Academic studies of the subject are contradictory, and areas of consensus are small. Work during high school seems to improve job prospects for those who don't plan on education after high school, but whether it helps or hurts the college-bound is less clear. And the question of whether widespread employment contributes to an overall loosening of academic standards is most elusive of all.

One thing that seems evident is that teenage work during secondary school is a particularly American phenomenon. A recent international survey of "middle-class" teenagers found that 60 percent of American teens work, about three times the average for all the other countries studied, and another found that Americans spend six times as many minutes per day at paying jobs as European counterparts. American teenagers also report more time doing chores for their families and, not surprisingly, less time reading and studying.

Well over half of American teenage workers toil in a distinct teenage economy of fast-food restaurants and retail stores. Such jobs give them little experience of working with adults, and because such jobs have been designed so that they require almost no manual, mathematical, or even social skills, they do not reinforce the young worker's schoolwork or point toward a future career. Those who support youth working note that such jobs nevertheless promote punctuality, adherence to standards, and a sense of autonomy. Those opposed note that schools are intended to inculcate these same habits, and that there is some evidence that heavy working lowers school attendance and academic ambition.

Despite parental beliefs that having a part-time job during high school will make their children better people, there is ample evidence that it is associated with behavior most parents don't endorse. Young people who work are more likely to use drugs, alcohol, and tobacco. The reasons seem obvious. They have the money that's required to indulge. Moreover, they are working like adults, and expect to win adult rewards, or at least assuage adult stress with adult addictions. Teenagers who work are also more likely to be sexually active.

The importance of teen spending is fairly well known. In 1995, a bad year for teen income, it still amounted to about $100 billion from wages and parental allowances, about half the size of the U.S. defense budget. The importance of teenage labor to the service economy is easily observed but little noticed. The low value of our young people's work helps give us affordable fast food and other services, just as the much lower cost of labor in Asia and Latin America makes possible inexpensive clothing and footware—much of which is purchased by teenagers with their small salaries. When, in 1996, legislation was introduced to raise the minimum wage, fast-food chains and other employers of teenagers were the most vehement opponents. They threatened massive layoffs of teenagers, and encouraged youth opposition to the bill. As it turned out, the legislation was passed in the midst of a tightening labor market, and the threatened layoffs never materialized.

★   ★   ★

Some of these disparate stories I've told about contemporary youth are about culture, others about public policy and law, still others about economics. Yet, they're all entangled with one another, and each illuminates a larger story—the difficulty of forging an identity in contemporary society.

As Erik Erikson described it at midcentury, the search for identity involves a period of experimentation as young people look at themselves and the society at large, and test and assess their talents and desires in relation to the world. Although he saw this process as a universal one, and used Martin Luther as a principal example, his analysis was probably time-bound, at least insofar as he assumed that it applied to people in all strata of society.

In previous centuries, relatively few young people have been offered the opportunity for the sort of psychosocial moratorium Erikson described. It may be that the only period of our history that offered the proper conditions for the overwhelming majority of young Americans to have what Erikson argued, and we still assume, to be the fundamental experience of adolescence were the decades from the end of World War II to the early 1970s. In the years since, we have extended the years of youthful irrelevance almost to thirty. But we have lessened young people's opportunities to test themselves, judge themselves, and find an identity that can last for the rest of their lives.

Today many teenagers know enough about adolescent psychology to speak the language of identity. When, as has recently happened throughout the country, school administrators propose that students be required to wear uniforms, students complain of being stripped of their identity.

But most of those things that are popularly called "identities" are simply categories, things that are, at most, parts of a fully formed identity. Being a member of a racial or ethnic group implies a set of shared experiences that help shape identities, but being an Italian-American, an African American, or a Hmong immigrant is not an identity in itself, no matter what we say. Neither, for that matter, is being gay or lesbian, or being wealthy, deaf, or dyslexic. You can have solidarity with others who share your experience or characteristics. But it's dangerous to accept so limited a conception of identity.

We tell young people how important it is for them to find their identity, yet offer them few avenues in which they can do so. Teenage tribalism is a kind of parody of identity, one that has existed for a very long time, but rarely so dramatically as in recent decades. Teenagers have left childhood behind and experience a surge of physical strength and sexuality, but our schools and employers show contempt for the talents and abilities of a significant minority of our young people. In attempting to encourage higher education, we have eliminated any path to maturity that involves actual making and doing. There is a youth employment ghetto, but no experience-based ladder of opportunity.

Because Americans have always expected to make their own fortunes, rather than inherit them, the issue of identity is inevitably associated with how you expect to make your living. The parents of many teenagers are still coping with this particular identity crisis. To some extent, these parents may be members of a generation that resists growing up. But the bigger reason is that the nature of work has changed. What Erikson called "occupational identity" is inevitably shaken in an environment where plants move overseas in search of cheaper labor, management structures are "flattened," professional staffs are "reengineered," and taking a chainsaw to the workforce is the surest way to make stock prices rise. I'm not arguing about whether such business methods are desirable or necessary. What I'm saying is that the very notion of psychological identity derives from a period when large, stable, intricately managed organizations demanded much from their workers, and were generally loyal to them in return. The jobs they offered may have been, as Paul Goodman argued, meaningless. But at least they were there. You didn't have to worry that you'd be fired, then faced with doing something even more stupid at lower pay.

As more and more years of schooling are required for even minimal participation in the economic mainstream, the chances increase that more and more young people will fall by the wayside. Except among Hispanics, high school dropout rates continue to fall and the educational aspirations of young people are continuing to increase. Young people can see what's going on. If they don't get enough schooling, they'll never even be considered for jobs that provide even the standard of living to which Americans are urged to aspire, much less the sense of pride and identity they are encouraged to have.

Schooling for young people in their teens has had two somewhat incompatible historical purposes. It has been, most of all, a sorting device, a means of maintaining class distinctions within an avowedly democratic society. And it has been, at least since the 1930s, a custodial institution, intended to delay entry into the labor force. These same two aims are evident today. The difference is that college, not high school, is the most important social screen. College has become more necessary, even as its expense has increased substantially. The graduate emerges, deeply in debt, to take a job that pays about as much as a high school graduate would have earned thirty years ago.

The declared reason for additional schooling is that tomorrow's jobs will require an array of technical, verbal, and human skills that require intensive preparation. However, many of the jobs expected to proliferate during the next two decades—caring for the elderly, for instance—will require knowledge and expertise, but not many years of schooling.

Moreover, many technical jobs, including very sophisticated computer-related tasks, are best learned through the discipline of trying to solve a real problem, rather than by memorizing abstract principles. During the high-tech employment boom of the mid-1990s, several top companies began recruiting people who hadn't even graduated from high school to work at the cutting edge of innovation. The young people were able to do the job. School and university are simply a convenient place to store them until their talents are required.

It's true, of course, that their skills will be obsolete in a few years. It is no longer realistic to assume that one's knowledge, or even the field you've studied, will last long enough to support you during your whole lifetime. Most people probably have to expect to spend their lives learning new things. But that's an argument against spending most of your youth in lengthy, narrowly focused preparation. In order to make sense, lifetime learning must accompany lifetime doing. Most of the time, though, we don't allow our young people to do much of anything.

Oddly, even as the workplace has become a less reliable arena in which to seek one's identity, respect for the nonoccupational bases for identity has weakened. It may have been chauvinist to assume, for example, that most girls would find their identities as wives and mothers. Yet, it's equally wrong to see something so fundamental as motherhood as a minor part of life that should be subordinated to what are, for most people, the dissatisfactions of work.

We look upon young unwed mothers as people who have made a poor career move, or as purchasers of a luxury item they cannot afford. Teenage motherhood is, in most cases, an impediment to developing a mature identity within the context of contemporary society. Yet, motherhood is as close to a meaningful way of spending one's life—as close to an identity, in other words—as most teenagers are able to achieve.

Having a baby is something a young woman can do that will transform her life, prod others to take her seriously, and win the attention of her family and the unconditional love of her baby. For the majority of young women, it's not a winning strategy in the long run. But it is a primally powerful act—one that's particularly attractive to young people whose experience has not encouraged them to look forward to a better life in the distant future. Motherhood gives those who lack abstract ambitions the chance to do something real.

Of all the cases of thwarted identity, partial identity, and false identity that afflict young people, perhaps the most confusing and damaging is also the one that appears most basic—that of the teenager.

As we've seen, teenagers bear an inordinate share of the blame for society's failures, while they're given too little responsibility for its improvement. Teenagers are people of whom too much is asked and too little is expected.

For various reasons, it has been rational, convenient, and even lucrative to consign young people to a different mode of living. But the young persist in wanting to do what their strong young bodies make them capable of doing: acting independently, working hard, having sex and families, and making lives—as they have during so much of this country's history.

A long historical perspective shows that young people have participated in American culture in a great variety of ways. The teenager was only one way. And now its time may be past. A teenager in the twenty-first century will appear as out of place as a goth in Tomorrowland.

FIFTEEN
Life After Teenagers

We must begin to think more clearly about teenagers, if only because, during the next few years, there will be a lot more of them.

During the first decade of the twenty-first century, the United States will have the largest number of teenagers in its history, more even than when the baby boomers bought their first blue jeans. The early years of this new century will, in large part, be shaped by this new generation, the largest infusion of youth in the U.S. population in more than four decades.

The number of people in their teens has been rising gradually for the last several years, reflecting the gradual rise in births that began in 1979. In the late 1980s, the birthrate shot up, and remained at more than 4 million births a year, the same number as during the peak years of the baby boom.

Those under eighteen now constitute 28 percent of the U.S. population, about 2 percent lower than at the height of the baby boom. Nor is this group as homogeneous as the boomers were; much of it is the second generation of the 1980s immigration wave. It won't turn us from an aging society to a youthful one, much less a predominantly under-twenty society like that found in some developing countries, and in the

United States a century and a half ago. Still, it's certain that so large a demographic wave as is moving through the U.S. population will make a difference.

This new wave of teens is growing up in a very different world than the boomers did. An economy driven by steel mills has given way to one based on microprocessors. The common experience of network television is replaced by the atomization of the Internet. The security of a good, steady job for a big, reliable company has given way to more precarious, and sometimes more rewarding, career paths. Nearly all the conditions that produced the classic teenager have changed, though we still assume that high school is the answer for everyone.

Like the baby boomers before them, this coming generation of youngsters made a big impact as tots. They asserted their market influence with their fanatical embrace of "The Mighty Morphin Power Rangers," both as toys and television. The enthusiasm for this odd, techno-mystical adventure series has long since cooled. But the society will forever after be divided between those who grew up knowing what a Zord is and those who'll never figure it out.

And like the teenage superheroes they made stars, the members of this Mighty Morphin generation seem to morph—to change their shape—each time you look. Everyone who studies this coming generation agrees that they're terribly important, and that their presence will change our society. But depending on who the observers are, what they're looking at, and what they expect to find, the coming teens appear to be monsters or saviors or anything in between.

Big spenders, for instance. Even though children and young people are the most impoverished age group in contemporary America, there's a large and growing literature on their spending habits. Teen-oriented clothing stores are opening in malls, and Wall Street has looked kindly on a number of new chains that serve this market. Teen-oriented publications are considered "hot" right now, after a long period when they were perceived as selling to a declining demographic.

Part of the attractiveness of this group for advertisers lies in the way in which they differ from the boomers. Busy parents in two-income households often have little time to buy groceries, for instance. Teenagers frequently do the shopping and help determine how to spend the family food budget.

The early word on them is that this new cohort isn't as depressive

or angry as their Gen-X grungy forebears and will be shapers and consumers of fashion, not simply recyclers. The theory is that, like the boomers, they have never lived through times of scarcity, and they feel
confident enough about their lives to be able to imagine ways of changing them for the better. Perhaps this is true, though it's not easy to know
what an eight-year-old will do at seventeen.

A lot of these new teens will be offspring of those who have arrived
in this country during the current immigration boom. The assimilation
of second-generation immigrants has, in the past, been a catalyst for
creativity—the second-generation Jews who helped create American pop
music, for instance.

On the surface, at least, youth culture has long been open and inclusive. Black urban teenagers are able, through their clothes, music, and
attitudes, to exert an influence on the society at large. Their parents,
grandparents, and even great-grandparents did the same. Generations of
white adults have muttered about jungle music and the dissipating effect
it would have on their young. Still, surveys indicate that today's young
people are much more open to dating across racial and ethnic lines than
their parents were (or are).

But these sunny views of a generation of responsible, creative, and
tolerant young morph into visions of a coming generation of barbarians,
raping and murdering their way into the millennium.

In most localities, especially major cities, the 1990s brought a dramatic decrease in crime, despite perceptions to the contrary. Larger numbers of young people probably will lead to an increase in crime; it almost
always does. Moreover, individual crimes by teenagers have become
steadily more violent.

Ever since these boomers were infants, law enforcement officials and
criminologists have been warning of the destructive potential of this generation. And even when crime has gone down, they have argued that
this is only the calm before the storm.

Then, in an instant, these murderous monsters morph into yet another kind of creature: teenagers "at-risk." A 1995 report by the Carnegie Council on Adolescent Development argued that half of all American
ten- to fourteen-year-olds are at risk of ruining their lives and blighting
society as a whole. "The problems have gotten worse," said David Ham-
burg, president of the Carnegie Corporation of New York when the
ten-year study was released. "Young teens engage in more and more

risky behavior. Things that used to be tried out in later adolescence are much more commonly occurring earlier—drugs, sex, and violence. The risks have gotten higher—from somewhat risky to very risky."

While most observers of "at-risk" youth point to the same complex of bad behavior, causes and solutions proposed depend largely on the politics of the observer. Those on the right fear nihilism and anarchy and blame bad parental values and fatherless households that are encouraged by federal welfare policies.

A more liberal perspective finds a steady disinvestment in the young during the last two decades, as spending on schools and children's health has suffered and only law enforcement and incarceration have been able to build a political consensus. Liberals see many of the same problems of poor educational achievement, suicide, drug use, sexually transmitted diseases, and violence, but they cite underfunded institutions, not bad morals.

Some leftists see teenagers as a postmodern proletariat, an easily manipulated, low-wage sector of the economy that helps keep all wages low, and many people out of the labor force altogether. The job of popular culture is to inure them to an economy that offers rewards for few and diversion for many.

This is, in fact, more or less the same thing some marketers are saying. A 1996 Dow Jones News Service article, reporting the results of the Rand Youth Poll, an annual marketing survey of teens, noted: "The transformation of American business to a service-driven one from manufacturing favors teen employment and is leading to higher personal spending by young people." For some, a generation of burger-flippers is a serious social problem; for others, it's a selling opportunity.

The one thing that all these visions of tomorrow's young share is that there will still be people called teenagers, and the assumptions society has made about them during the last sixty years or so will continue to hold true. Among these are that teenagers are immature and thus dependent on their families, that they should be protected from the world of work, and that their primary responsibility should be to go to school to prepare themselves for a lifetime of employment.

We have seen that many of these assumptions have been false for decades. Teenagers are not protected. When they are home, they're often alone there. They are more likely to miss a day of school than a day of work. And the idea of training in youth for a life's work is increasingly obsolete.

By looking at history as we have done, we see that the teenager that emerged in the mid-twentieth century was but one of many ways in which American young people have responded to the circumstances of their times.

Throughout history, most young Americans have not led sheltered lives of study and preparation. They have supported their families as they struggled to survive. They have been pioneers and entrepreneurs. They have been poor or displaced, left to scramble on their own as bootblacks or newsboys, or as pickpockets and prostitutes. They've been soldiers and sailors and cowboys and miners and schoolteachers and physicians. At most times, only a few have been students, living at home, devoting their second decade to preparing for the future.

It is this last indulged group that contributed most to the classic conception of the teenager. This mode of being young emerged when work moved out of the home, when powerful families felt themselves buffeted by change, and some women traded an economic role for moral authority. This model of the bourgeois home has had a pretty long run—more than 150 years thus far—though only during the last sixty have we assumed that it could be and ought to be universal. This seemed to work for a while, during the aftermath of World War II, provided that you overlooked "the other America"—nearly all minority groups and members of the urban and rural poor. For the last twenty-five years, even those who have grown up expecting to live normal, middle-class lives have had to strain to do so.

Today's young people pass through metal detectors at the doors of their high schools. They come home to houses that are, by historical standards, large and well equipped, but which both parents are working hard to pay for. Often, they work long hours themselves.

And in poor neighborhoods, national gangs like the Bloods—whose colors and brand identity are found in nearly as many cities as McDonald's—recruit aggressively, offering high-risk, high-reward opportunities.

Some young people have serious problems, and statistics suggest that they are worse and more widespread than they were a generation ago. Young people have become more violent, more prone to suicide, more sexually active, more likely to be victims of crime, more prone to drug and alcohol abuse, and in every way both more dangerous and more endangered. (Several of these indicators took a turn for the better in the mid-1990s, though nobody feels confident that the improvements are permanent.)

The great majority of young people grow up all right, but even they are affected by the stigma and atmosphere of mistrust that has developed in response to the negative trends among the young. Indeed, the normal, law-abiding young, facing pervasive suspicion in their everyday lives and increasing restrictions on their freedom of movement and association, are paying a high price for problems that aren't theirs.

Young people are exposed to all the violence and economic insecurity of the society at large, but, unlike their predecessors, they have few avenues for bearing real responsibility to improve their situation. We make it difficult for teenagers to imagine themselves living useful lives. They are offered few immediate and meaningful ways to test their new-found powers, to feel needed, to be essential members of a community. We tell them that they're the future—and it's true—but they need opportunities to take responsibility for the here and now.

If we are going to persist in our notion that all young people should spend their teens simply waiting for adulthood, we have to find ways of making this teenage experience more satisfying and effective.

Alternatively, we can decide that the idea of the teenager is one that has outlived its usefulness, then move on to other possibilities.

Or—and I believe that this is the right answer—we can do both.

By that I mean that we can accept that being dependent on one's parents, going to school, and preparing for a distant future is a good and appropriate way for many young people to spend their lives. Thus, we can pursue ways to make high schools better, to better coordinate work opportunities with education, and to encourage student volunteerism without destroying its essence by making it mandatory. And parents can consider how to be a greater presence in the lives of their children.

At the same time, we can recognize that many people are better suited to follow other life courses. They are people with personalities and talents that would be better served by making working and learning more flexible, by removing the stigma of the school dropout and increasing opportunities to be school drop-ins.

My point is not to assert all people in their teens should be considered full-fledged adults and be treated accordingly. Rather, I'm arguing that they should be treated as beginners—inexperienced people who aren't fundamentally different from adults, but who, because they are dealing with so many new things in their lives, usually need more help, more attention, and more patience than those who have more experience.

In other words, we need to get rid of G. S. Hall's discredited notion of the adolescent as incompetent, troubled, half-mad, and dangerous, along with the stereotype of "raging hormones" that gives that old prejudice a pseudoscientific veneer. But we needn't get rid of adolescence as Erik Erikson described it, as a period of learning, observation, experimentation, and identity-seeking. Indeed, one of the problems with the contemporary notion of the teenager is that it impedes this Eriksonian adolescent struggle and channels its risk-taking into areas that are pointless or self-destructive.

Youth should be a time for learning that one's decisions have consequences—although not necessarily irreversible ones. Young people should be encouraged to experiment. They should be allowed to leave school for a while without stigma to learn something about the world of work. They have should the opportunity to try something new and unlikely—and to fail at it—without being branded a failure for life.

By providing so few alternatives for young lives, we increase the chances that young people will fail in big ways—by becoming teenage mothers or dropouts with few economic prospects. We need to find ways to let them fail in small ways so that they can learn something each time they do so.

Young people ought to be able to follow many different paths, yet be protected—as beginners—from going too far astray. (It works on ski slopes.) These are many different ways of living your life. Why should we pretend that there is only one way to be young?

My point in writing this book has been to offer perspectives, not prescriptions. Many of the sorts of programs and policies that might help realize this vision of life after teenagers already exist, at least on a local level.

For example, the charter school movement offers at least the possibility that schooling can be organized around a specific goal, other than that of using up young people's time. Others advocate judging, and graduating, students, based on an assessment of what they've really learned, rather than simply on time served. Meanwhile, some corporations have programs that allow workers to integrate their schooling with what they need in their jobs.

The intensive probation programs like those in Boston, discussed in the previous chapter, point the way toward making young people feel less like part of an anonymous mass. Those who are not in trouble with the law might be able to benefit from having someone who can help them navigate through an otherwise bewildering array of educational and occupational choices.

The possibilities are close to endless. What they have in common is that they would deal with teenagers not as a huge and monolithic population, but as people who are just starting out. They will make their own lives, but they will need some help along the way.

Today's young people are heirs to all the different possibilities for youth that their predecessors have embodied: the fourteen-year-old Puritan searching anxiously for a calling; the Plains Indian starving himself in a quest for a vision; the bookish, exhausted factory girls of Lowell; the foul-mouthed, whiskey-swilling, barely pubescent go-getters of the Rocky Mountain frontier; and all the rest. Most of these earlier young Americans aren't role models. In fact, if they turned up today, some would be considered troubled teens. Many of them faced conditions that seem barbaric today, though they in turn might be horrified by the violence and insecurity with which many contemporary youths live.

All those young people in our country's past, like today's youth, had to make do with the world as they found it. We speak carelessly of the idealism of youth, but what young people need and often exhibit is a fierce realism. They need to get beyond the things we tell them and figure out how things really work. And when they act on what they know, the results sometimes frighten their elders. The job of young people is not, as we sometimes assume, simply to go to high school. It is to imagine and begin to construct their lives. They need to understand both their own interests and abilities and the society of which they are part. And they need to make a self that makes sense for the times in which they live.

This isn't easy. It never has been. It never will be.

And it's not easy for the elders to see their ideas supplanted and their hard-earned wisdom casually ignored.

Yet, we are the ones who create the conditions in which young people face the task of creating their lives. We erect the thresholds. We tell them much of what they know. Our own fears and biases color the way they see the world. Despite widespread suspicions to the contrary, young people believe most of what their parents and other adults in authority tell them. One of the most important, confusing, and even damaging things we tell them is that they are teenagers.

Once we understand that the teenager—this weird, alienated, frightening yet enviable creature—is a figment of our collective imagination, the monstrous progeny of marketing and high school, all generations will benefit. Young people will be freer to define themselves as individuals, yet more closely tied to the families and communities of which they are a part.

We will be able to see the young as people—beginners, to be sure, but people nonetheless. Their ills won't vanish instantly, but it will be easier for people of all generations to understand the role they play in creating the problems and the stake they have in solving them. The young will still demand attention, but it can't be the kind of condescending attention we might give to a brute with raging hormones who's merely passing through a phase.

Nobody grows out of being a person, and the second decade of life is when one's personhood becomes defined in terms of the wider world. As a society, we need to support that process, not sequester and render it irrelevant.

Rather than try to force young people to delay using their newfound powers, we should accept youth as a gift that can benefit us all. In a rapidly aging society, youth may prove to be a precious resource, a source of energy to revivify a culture addicted to reruns.

Young people are here and, as always, ready to make history. Let's help them. We were young once too.

SOURCES AND FURTHER READING

While the primary concern of this book is to describe the many ways in which young lives have changed in the last four centuries, my research has inevitably focused my attention on changing adult concerns about contemporary teenagers. For example, when I began my research in 1995, there was a lively controversy as to whether an increased minimum wage would throw teenagers out of work. The minimum wage was increased without ill effects on young people, largely because overall employment was growing. Indeed, as the book neared completion in 1998, the worry was that teenagers were working too much, that wages were becoming too high, and that high-tech companies were tempting young people to forgo college and even high school. The anxiety level remained the same; what people worried about changed completely.

I soon realized that it would be dangerous for the book to be preoccupied with the latest crisis or the latest study, because both would be old news by the time you read this. Yet, the book is nonetheless shaped by the issues and information about contemporary teenagers that emerged while I was researching and writing it. This was the first book that I have written making use of the Internet. Every morning the service My Excite presented me with two dozen or so articles from newspapers and magazines around the world dealing with youth, their problems, and their culture. I thought that about 1,500 of them were interesting enough to save, and a few made their way into the book, though not, for the most part, into this bibliography. The Internet is particularly useful for obtaining exhaustive local coverage of an event. Thus, my account of the young woman accused of murdering her baby at the senior prom uses accounts from the *Asbury Park Press* and reaction from around the country.

Statistical data on young people is tricky to work with because nearly everyone who compiles data divides up the adolescent age group differently. Most of the statistics about the current state of American teenagers in this book rely on the most recent available versions of the U.S. Department of Education's *Youth Indicators* (see Snyder below); "Add Health," a federally funded long-term national study of young people that defines health in very broad terms (see Resnick below); the biennial *Kids Count* surveys of the Anne E. Casey Foundation; and the various reports of the Carnegie Commission on Adolescent Development. By the time you read this, most of these institutions will have new findings and more up-to-date statistics.

Some sources are inevitably more important than others. Nobody can write on this subject without referring to the work of Joseph F. Kett. I found the work of Harvey Graff to be valuable, and I relied particularly on the fine anthology he edited. Chapter 10 is strongly influenced by the work of John Modell and Kathy Lee Peiss.

The single most important source for this book is undoubtedly the massive five-volume work *Children and Youth in America: A Documentary History,* edited by Robert H. Bremner, with John Barnard, Tamara K. Hareven, and Robert M. Mennel. This compilation of original documents has a bureaucratic bias, but it is by far the best thing of its kind available. Another tremendously valuable source of original documents is the Depression-era Work Projects Administration's Life Histories Collection of the Library of Congress, which is available at the library's web site (www.LOC.gov).

The list that follows is selective. I have included only those books and articles on which I relied strongly, or which I think would be most useful to the reader who wishes to explore further.

Adams, Samuel Hopkins ["Warner Fabian"]. *Flaming Youth*. New York: Boni and Liveright, 1923.

Adelson, Joseph. *Inventing Adolescence*. New Brunswick, N.J.: Transaction, Inc., 1986.

Alcott, Louisa May. *Little Women*. Boston: Little, Brown, 1968 [1868].

Alexander, Ruth M. " 'The Only Thing I Wanted Was Freedom': Wayward Girls in New York, 1900–1930." In West and Petrik (eds).

Alger, Horatio, Jr. *The Cash Boy*. New York: A. L. Burt, 1887.

———. *Ragged Dick; or, Street Life in New York with the Bootblacks*. Boston: Loring, 1868.

Ann E. Casey Foundation. *1997 Kids Count, Summary and Findings*. Baltimore: Ann E. Casey Foundation, 1997.

Ariés, Philippe. *Centuries of Childhood*. (Trans. Robert Baldick.) New York: Vintage, 1962.

Arnett, Jeffrey Jensen. *Metalheads: Heavy Metal Music and Adolescent Alienation*. Boulder, Colo.: Westview Press, 1996.

Bailey, Beth L. *From Front Porch to Back Seat: Courtship in Twentieth-Century America*. Baltimore: Johns Hopkins University Press, 1988.

Bakan, David. "Adolescence in America: From Idea to Social Fact." [1971] In Robert E. Grinder (ed.), *Studies in Adolescence*. (Third Edition.) New York: Macmillan, 1975.

Beales, Ross W., Jr. "In Search of the Historical Child: Miniature Adulthood and Youth in Colonial New England." [1975] In Harvey T. Graff (ed.), *Growing Up in America: Historical Experiences*. Detroit: Wayne State University Press, 1987.

Benedict, Ruth. *Patterns of Culture*. Boston: Houghton Mifflin, 1934.

Bettelheim, Bruno. *Symbolic Wounds: Puberty Rites and the Envious Male*. (Revised Edition.) New York: Collier Books, 1962.

Block, Herbert A., and Arthur Niederhoffer. *The Gang: A Study in Adolescent Behavior*. New York: Philosophical Library, 1958.

Blos, Peter. *On Adolescence: A Psychoanalytic Interpretation*. New York: Free Press, 1958.

Blumberg, Joan Jacobs. "Chlorotic Girls, 1870–1920: A Historical Perspective on Female Adolescence." *Child Development*, Vol. 53, December 1982. Chicago: University of Chicago Press.

———. *The Body Project*. New York: Random House, 1997.

Blumer, Herbert. "Motion Picture Autobiographies." [1933] Excerpted in Robert Sklar (ed.), *The Plastic Age: 1917–1930*. New York: George Braziller, 1970.

Bodnar, John E. "Socialization and Adaptation: Immigrant Families in Scranton, 1880–1890." [1976] In Graff (ed.), 1987.

Bremner, Robert H. (ed.) *Children and Youth in America: A Documentary History*. Cambridge, Mass.: Harvard University Press, 1970.

Bremner, Robert H. "Rights of Children and Youth." In James S. Cole-
    man (ed.) *Youth: Transition to Adulthood. Report of the Panel on Youth
    of the President's Science Advisory Committee.* Chicago: University of
    Chicago Press, 1974.

Brown, Frederic Kenyon ["Al Priddy"]. *Through the Mill: The Life of a
    Mill Boy.* Boston: Pilgrim Press, 1911.

Carnegie Council on Adolescent Development. *Great Transitions—Prepar-
    ing Adolescents for a New Century.* New York: Carnegie Council on
    Adolescent Development, 1996.

Cohen, Albert K. *Delinquent Boys.* New York: Free Press, 1955.

Coleman, James C. (ed.). *Transition to Adulthood: Report of the Panel on
    Youth of the President's Science Advisory Committee.* Chicago: University
    of Chicago Press, 1974.

————. "Youth Culture." In Coleman (ed.), 1974.

Coleman, James S. *The Adolescent Society.* New York: Free Press, 1961.

Conant, James Bryant. *The American High School Today: A First Report to
    Interested Citizens.* New York: McGraw-Hill, 1959.

————. *The Revolutionary Transformation of the American High School.*
    Cambridge, Mass.: Harvard University Press, 1959.

Coté, James E., and Anton L. Allahar. *Generation on Hold: Coming of
    Age in the Late Twentieth Century.* New York: New York University
    Press, 1996.

Csikszentmihalyi, Mihaly, and Reed Larson. *Being Adolescent: Conflict and
    Growth in the Teenage Years.* New York: Basic Books, 1984.

Davis, Maxine. *The Lost Generation: A Portrait of America's Youth Today.*
    New York: Macmillan, 1936.

Demos, John. "The Rise and Fall of Adolescence." In *Past Present and
    Personal.* New York: Oxford University Press, 1986.

————. *The Unredeemed Captive.* New York: Alfred A. Knopf, 1994.

Divoky, Diane (ed.). *How Old Will You Be in 1984? Expressions of Student
    Outrage from the High School Free Press.* New York: Discus Books, 1969.

Dryfoos, Joy G. *Adolescents at Risk: Prevalence and Prevention.* New York:
    Oxford University Press, 1990.

Edmonds, Franklin Spencer. *History of the Central High School of Philadel-
    phia.* Philadelphia: Lippincott & Co., 1902.

Eichorn, Dorothy. "Biological, Psychological, and Socio-Cultural Aspects
    of Adolescence and Youth." In Coleman (ed.), 1974.

Eisenstadt, S. N. "Archetypal Patterns of Youth." [1961] In Graff (ed.),
    1987.

————. *From Generation to Generation: Age Groups and Social Structure.* Glencoe, Ill.: Free Press, 1956.

Elder, Glen H., Jr. "Adolescence in Historical Perspective." [1980] In Graff (ed.), 1987.

Erikson, Erik H. *Childhood and Society.* New York: W. W. Norton, 1950.

————. *Identity, Youth, and Crisis.* New York: W. W. Norton, 1968.

————. *Young Man Luther: A Study in Psychoanalysis and History.* New York: W. W. Norton, 1958.

Farrell, Edwin William. *Hanging In and Dropping Out: Voices of At-Risk High School Students.* New York: Teachers College Press, 1990.

Fass, Paula S. *The Damned and the Beautiful: American Youth in the 1920s.* New York: Oxford University Press, 1977.

Fitzgerald, F. Scott, *This Side of Paradise.* New York: Scribner's, 1920.

————. *Echoes of the Jazz Age.* New York: Scribner's, 1931.

Fox, Vivian C. "Is Adolescence a Phenomenon of Modern Times?" *Journal of Psychohistory.* Vol. 5, No. 2, 1977. New York: Association for Psychohistory.

Freud, Anna. "Adolescence," *Psychoanalytic Studies of the Child.* New York: International Universities Press, 1958.

Friedenberg, Edgar Z. *The Vanishing Adolescent.* Boston: Beacon Press, 1959.

————. *Coming of Age in America.* New York: Random House, 1965.

Gaines, Donna. *Teenage Wasteland: Suburbia's Dead End Kids.* New York: Pantheon, 1991.

Gilbert, James B. *A Cycle of Outrage: America's Reaction to the Juvenile Delinquent of the 1950s.* New York: Oxford University Press, 1986.

Gilbert, Marc Jason. "Lock and Load High." *Vietnam Generation Journal.* Vol. 5, No. 1–4, 1994. Charlottesville, Va.: Vietnam Generation, 1994.

Goodman, Paul. *Growing Up Absurd.* New York: Random House, 1960.

Graebner, William. *Coming of Age in Buffalo: Youth and Authority in the Postwar Era.* Philadelphia: Temple University Press, 1990.

————. "Outlawing Teenage Populism: The Campaign Against Secret Societies in the American High School, 1900–1960." *Journal of American History.* Vol. 74, No. 2, September 1987. Bloomington, Ind.: Organization of American Historians.

Graff, Harvey J. *Conflicting Paths: Growing Up in America.* Cambridge, Mass.: Harvard University Press, 1995.

Graff, Harvey J. (ed.). *Growing Up in America: Historical Experiences.* Detroit: Wayne State University Press, 1987.

Greenberger, Ellen, and Laurence Steinberg. *When Teenagers Work: The*

Psychological and Social Costs of Adolescent Employment. New York: Basic Books, 1986.

Greven, Philip J. *Four Generations: Population, Land, and Family in Colonial Andover, Massachusetts*. Ithaca, N.Y.: Cornell University Press, 1970.

———. "Youth, Maturity, and Religious Conversion: A Note on the Ages of Converts in Andover, Massachusetts, 1711–1749." [1972] In Graff (ed.), 1987.

Griliches, Zvi. "Economic Problems of Youth." In Coleman (ed.), 1974.

Grinder, Robert E. (ed.). *Studies in Adolescence*. New York: Macmillan, 1975.

Gross, Ronald, and Paul Osterman (eds.). *High School*. New York: Simon & Schuster, 1972.

Grubb, Norton, and Marvin Lazerson. *Broken Promises: How Americans Fail Their Children*. Chicago: University of Chicago Press, 1982.

Gutman, Herbert G. *The Black Family in Slavery and Freedom, 1750–1925*. New York: Pantheon, 1976.

Hall, Granville Stanley. *Adolescence: Its Psychology and Its Relations to Physiology, Anthropology, Sociology, Sex, Crime, Religion, and Education*. New York: D. Appleton and Company, 1904.

———. "Flapper Americana Novissima." *The Atlantic Monthly*. Boston, June 1922.

Hart, Richard L., and J. Galen Saylor (eds.). *Student Unrest: Threat or Promise?* Washington, D.C.: Association for Supervision and Curriculum Development, 1970.

Hollingshead, August B. *Elmtown's Youth and Elmtown Revisited*. New York: John Wiley, 1974 [1949].

Ianni, Francis A. J. *The Search for Structure: A Report on American Youth Today*. New York: Free Press, 1989.

Jones, Landon Y. *Great Expectations: America and the Baby Boom Generation*. New York: Coward, McCann and Geoghegan, 1980.

Kantor, Harvey. "Managing the Transition from School to Work: The False Promise of Youth Apprenticeship." *Teachers College Record*. June 1, 1994. New York: Teachers College of Columbia University.

Keniston, Kenneth. "Psychological Development and Historical Change." [1971] In Graff (ed.), 1987.

Kett, Joseph F. *Rites of Passage: Adolescence in America, 1790 to the Present*. New York: Basic Books, 1977.

———. "Growing Up in Rural New England 1800–1840." [1971] In Graff (ed.), 1987.

Kiell, Norman, *The Universal Experience of Adolescence*. New York: International Universities Press, 1964.

Kleijwegt, Marc. *Ancient Youth: The Ambiguity of Youth and the Absence of Adolescence in Greco-Roman Society*. Amsterdam: J. C. Gieben, 1991.

Krug, Edward A. *The Shaping of the American High School*. New York: Harper & Row, 1964.

———. *The Shaping of the American High School, 1920–1941*. Madison, Wisc.: University of Wisconsin Press, 1972.

Larabee, David F. *The Making of an American High School: The Credentials Market and the Central High School of Philadelphia, 1838–1939*. New Haven: Yale University Press, 1988.

Larkin, Ralph W. *Suburban Youth in Cultural Crisis*. New York: Oxford University Press, 1979.

Lefkowitz, Bernard, *Tough Change: Growing Up on Your Own in America*. New York: Free Press, 1987.

Libarle, Marc, and Tom Seligson (eds.). *The High School Revolutionaries*. New York: Random House, 1970.

Lindley, Betty (Grimes), and Ernest K. Lindley. *A New Deal for Youth: The Story of the National Youth Administration*. New York: Viking Press, 1938.

Luker, Kristin. *Dubious Conceptions: The Politics of Teenage Pregnancy*. Cambridge, Mass.: Harvard University Press, 1996.

Lynd, Robert S., and Helen Merrill Lynd. *Middletown: A Study in Contemporary American Culture*. New York: Harcourt, Brace and Co., 1929.

———. *Middletown in Transition*. New York: Harcourt, Brace and Co., 1937.

Macleod, David I. "Act Your Age: Boyhood, Adolescence, and the Rise of the Boy Scouts of America." *Journal of Social History*. 1982. Pittsburgh: Carnegie-Mellon University Press.

Males, Mike A., *The Scapegoat Generation*. Monroe, Me.: Common Courage Press, 1996.

Mallery, David. *High School Students Speak Out*. New York: Harper & Brothers, 1961.

Maynard, Rebecca A. (ed.). *Kids Having Kids: Economic Costs and Social Consequences of Teen Pregnancy*. Washington, D.C.: Urban Institute Press, 1997.

Mead, Margaret. *Coming of Age in Samoa*. New York: William Morrow, 1928.

Meyrowitz, Joshua. "The Adultlike Child and the Childlike Adult: Socialization in an Electronic Age." [1984] In Graff (ed.), 1987.

Mirel, Jeffrey. "From Student Control to Institutional Control of High School Athletics: Three Michigan Cities, 1883–1905." *Journal of Social History,* 1982. Pittsburgh: Carnegie-Mellon University Press.

Modell, John. *Into One's Own: From Youth to Adulthood in the United States, 1920–1975.* Berkeley, Calif.: University of California Press, 1989.

Modell, John, Frank F. Furstenberg, Jr., and Theodore Hershberg. "Transitions to Adulthood." [1976] In Graff (ed.), 1987.

Moffatt, Michael. *Coming of Age in New Jersey.* New Brunswick, N.J.: Rutgers University Press, 1989.

Morgan, Edmund S. *The Puritan Family: Essays on Religion and Domestic Relations in Seventeenth-Century New England.* Boston: Trustees of the Public Library, 1944.

Muir, John. *My Boyhood and Youth.* Boston: Houghton Mifflin, 1913.

Musto, David F. "Opium, Cocaine, and Marijuana in American History." *Scientific American.* July 1991.

Nasaw, David. "Children and Commercial Culture: Moving Pictures in the Early Twentieth Century." In West and Petrik, 1992.

Nathanson, Constance. *A Dangerous Passage: The Social Control of Sexuality in Women's Adolescence.* Philadelphia: Temple University Press, 1991.

Norris, Clarence, and Sybil B. Washington. *The Last of the Scottsboro Boys.* New York: G. P. Putnam's Sons, 1979.

Odem, Mary E. *Delinquent Daughters: Protecting and Policing Adolescent Female Sexuality in the United States, 1885–1920.* Chapel Hill, N.C.: University of North Carolina Press, 1995.

Offer, Daniel, Eric Ostrov, and Kenneth I. Howard. *The Adolescent: A Psychological Self-Portrait.* New York: Basic Books, 1981.

Offer, Daniel (with Melvin Sabshin, and Judith L. Offer). *The Psychological World of the Teenager: A Study of Normal Adolescent Boys.* New York: Basic Books, 1969.

Offer, Daniel, and Judith Baskin Offer (with Eric Ostrov). *From Teenage to Young Manhood: A Psychological Study.* New York: Basic Books, 1975.

Palladino, Grace. *Teenagers: An American History.* New York: Basic Books, 1996.

Parsons, Talcott. "Age and Sex in the Social Structure of the United

States." [1942] In Parsons, *Essays in Sociological Theory.* Revised Edition. Glencoe, Ill.: Free Press, 1954.

Peiss, Kathy Lee. *Cheap Amusements: Working Women and Leisure in New York City, 1880 to 1920.* Philadelphia: Temple University Press, 1986.

Pipher, Mary Bray. *Reviving Ophelia: Saving the Selves of Adolescent Girls.* New York: Putnam, 1994.

Radin, Paul. *The World of Primitive Man.* New York: Grove Press, 1953.

Reese, William J. *The Origins of the American High School.* New Haven: Yale University Press, 1995.

Reiman, Richard A. *The New Deal and American Youth.* Athens, Ga.: University of Georgia Press, 1992.

Resnick, Michael D., et al. "Protecting Adolescents from Harm: Findings of the National Longitudinal Study on Adolescent Health (Add Health)." *Journal of the American Medical Association.* September 10, 1997. Chicago: American Medical Association.

Roth, Randolph. "Wayward Youths: Raising Adolescents in Vermont, 1777–1815." *Vermont History.* Vol. 59, No. 2, 1991. Montpelier, Vt.: Vermont Historical Society.

Rousseau, Jean-Jacques. *Emile, or On Education.* (Trans. Allan Bloom.) New York: Basic Books, 1979.

Rubin, Nancy J. *Ask Me If I Care: Voices from an American High School.* Berkeley, Calif.: Ten Speed Press, 1994.

Ryan, Mary P. "Privacy and the Making of the Self-Made Man: Family Strategies of the Middle Class at Midcentury." [1981] In Graff (ed.), 1987.

Ryder, Norman. "The Demography of Youth." In Coleman (ed.), 1974.

Schultz, James A. "Medieval Adolescence: The Claims of History and the Science of German Narrative." *Speculum.* Vol. 66, No. 3, July 1991. Cambridge, Mass.: Medieval Academy of America.

Schwartz, Gary, and Don Merten. "The Language of Adolescence: An Anthropological Approach to Youth Culture." [1967] In Grinder (ed.), 1975.

Scott, Donald M., and Bernard Wishy (eds.). *America's Families: A Documentary History.* New York: Harper & Row, 1982.

Selden, Bernice. *The Mill Girls.* New York: Atheneum, 1983.

Sizer, Theodore R. *Horace's Compromise: The Dilemma of the American High School.* Boston: Houghton Mifflin, 1984.

————. *Horace's Hope: What Works for the American High School.* Boston: Houghton Mifflin, 1996.

Smith, Daniel Scott. "Parental Power and Marriage Patterns: An Analysis of Historical Trends in Hingham, Mass." [1973] In Graff (ed.), 1987.

Smith, Ernest A. *American Youth Culture: Group Life in Teenage Society.* New York: Free Press of Glencoe, 1962.

Snyder, Thomas (ed.). *Youth Indicators, 1996.* Washington, D.C.: National Center for Educational Statistics, U.S. Department of Education, 1996.

Spacks, Patricia Meyer. *The Adolescent Idea.* New York: Basic Books: 1981.

Stansell, Christine. "Women, Children, and the Use of the Streets: Class and Gender Conflict in New York City, 1850–1860." [1982] In Graff (ed.), 1987.

Tarkington, Booth. *Seventeen: A Tale of Youth and Summer Time and the Baxter Family, Especially William.* New York: Harper & Brothers, 1916.

Thompson, Sharon. *Going All the Way: Teenage Girls' Tales of Sex, Romance, and Pregnancy.* New York: Hill and Wang, 1995.

Ueda, Reed. *Avenues to Adulthood: The Origins of the High School and Social Mobility in an American Suburb.* New York: Cambridge University Press, 1987.

Vallone, Lynne. *Disciplines of Virtue: Girls' Culture in the Eighteenth and Nineteenth Centuries.* New Haven: Yale University Press, 1995.

Van Gennep, Arnold. *The Rites of Passage.* (Trans. Monika B. Vizedom and Gabrielle L. Caffee.) Chicago: University of Chicago Press, 1960.

Vinovskis, Maris. *An "Epidemic" of Adolescent Pregnancy? Some Historical and Policy Considerations.* New York: Oxford University Press, 1988.

Walsh, Lorena S. " 'Till Death Us Do Part': Marriage and Family in Seventeenth-Century Maryland." [1979] In Graff (ed.), 1987.

Webber, Thomas L. "The Setting: Growing Up in a Quarter Community." [1978] In Graff (ed.), 1987.

West, Elliott. "Heathens and Angels: Childhood in the Rocky Mountain Mining Towns." [1983] In Graff (ed.), 1987.

West, Elliott, and Paula Petrik (ed.). *Small Worlds: Children and Adolescents in America, 1850–1950.* Lawrence, Kan.: University Press of Kansas, 1992.

Zall, P. M. *Becoming American: Young People in the American Revolution.* Hamden, Ct.: Linnet Books, 1993.

INDEX